AF892537

re information about this series at https://link.springer.com/bookseries/4604

Engineering Materials and Process

Series Editor

Brian Derby, Materials Science Center, University of Manchester,

M

Antonios Kanellopoulos ·
Jose Norambuena-Contreras
Editors

Self-Healing Construction Materials

Fundamentals, Monitoring and Large Scale Applications

Springer

Editors
Antonios Kanellopoulos
Division of Civil Engineering and Built Environment, School of Physics Engineering and Computer Science
University of Hertfordshire
Hatfield, Hertfordshire, UK

Jose Norambuena-Contreras
LabMAT, Ingeniería Civil Ambiental
Department of Civil and Environmental Engineering
Universidad del Bío-Bío
Concepcion, Chile

ISSN 1619-0181 ISSN 2365-0761 (electronic)
Engineering Materials and Processes
ISBN 978-3-030-86879-6 ISBN 978-3-030-86880-2 (eBook)
https://doi.org/10.1007/978-3-030-86880-2

© Springer Nature Switzerland AG 2022
This work is subject to copyright. All rights are reserved by the Publisher, whether the whole or part of the material is concerned, specifically the rights of translation, reprinting, reuse of illustrations, recitation, broadcasting, reproduction on microfilms or in any other physical way, and transmission or information storage and retrieval, electronic adaptation, computer software, or by similar or dissimilar methodology now known or hereafter developed.
The use of general descriptive names, registered names, trademarks, service marks, etc. in this publication does not imply, even in the absence of a specific statement, that such names are exempt from the relevant protective laws and regulations and therefore free for general use.
The publisher, the authors and the editors are safe to assume that the advice and information in this book are believed to be true and accurate at the date of publication. Neither the publisher nor the authors or the editors give a warranty, expressed or implied, with respect to the material contained herein or for any errors or omissions that may have been made. The publisher remains neutral with regard to jurisdictional claims in published maps and institutional affiliations.

This Springer imprint is published by the registered company Springer Nature Switzerland AG
The registered company address is: Gewerbestrasse 11, 6330 Cham, Switzerland

Preface

Maintaining structural integrity is of paramount importance for the reliability of civil infrastructure. Over the years, the construction sector has developed the reputation of being conservative and rigid in adopting change and innovation. In addition, for many decades the construction materials were thought of as large volume consumables that, when had to be replaced, they would be replaced. Given the vastness of the construction sector globally, it is apparent that this approach has significant environmental consequences and is not sustainable. The life of civil engineering materials and their components must be extended to ensure resilient structural systems. On technical level, a major misconception in civil engineering practice is that our construction materials are robust. Even design standards, such as the Eurocode 2 on the design of reinforced concrete structures, consider cracking of concrete up to 300 μm not a big issue. Indeed, such cracks do not pose an immediate structural threat, but they are the locations from where degrading, to the steel reinforcement, compounds will penetrate. Asphalt and steel are facing similar issues. Extensive maintenance regimes are typically put in place to alleviate degradation arising from in-service actions. Nonetheless, the maintenance regimes do not always rectify the situation. Microcracks can remain in the system and continue contributing to the reduction of the overall integrity of a structure.

Self-healing materials have attracted great interest in the last two decades due to their several applications for civil engineering. A self-healing material is defined as an artificial or synthetically created material that has the built-in ability to repair damage to itself without human intervention. For self-healing construction materials, the last decade has been characterised with a considerable amount of published work in the field. A growing number of research groups and academics worldwide have been involved in developing self-healing concepts for the main civil engineering materials including concrete, steel and asphalt. The reader must keep in mind that this book overviews the existing state of the art in a very dynamic and fast-evolving scientific field. It includes information about experimental work, numerical simulations and large-scale trials. Its last chapter highlights the future directions and the emerging principles in the area of smart-adaptive construction materials. The following chapters intend to elucidate the concepts and mechanisms of self-healing construction materials to practising engineers, researchers and students.

Hatfield, UK — Dr. Antonios Kanellopoulos
Concepcion, Chile — Dr. Jose Norambuena-Contreras

Acknowledgements The editors would like to express their gratitude to all the contributing authors of this book. Their efforts and dedicated work have been a key element for the success of this project, developed majorly during the global COVID-19 pandemic.

Contents

Fundamentals of Self-healing Construction Materials 1
Antonios Kanellopoulos and Jose Norambuena-Contreras

Self-Healing Cement-Based Materials: Mechanisms and Assessment 13
Antonios Kanellopoulos

Self-Healing in Metal-Based Systems 43
Mariia Arseenko, Julie Gheysen, Florent Hannard, Nicolas Nothomb, and Aude Simar

Advances in Self-healing Bituminous Materials: From Concept to Large-Scale Application 79
Jose Norambuena-Contreras, Quantao Liu, Alvaro Gonzalez, Alvaro Guarin, Nilo Ruiz-Riancho, Alvaro Garcia-Hernandez, Bastian Wacker, and Jose L. Concha

Multiscale Measurements of the Self-Healing Capability on Bituminous Materials 127
Guoqiang Sun, Daquan Sun, Alvaro Guarin, and Jose Norambuena-Contreras

Numerical Simulation of Self-Healing Cementitious Materials 151
B. L. Freeman and A. D. Jefferson

Modeling of Self-healing Process in Bituminous Materials: Experimental and Numerical Models 187
Guoqiang Sun, Daquan Sun, Mingjun Hu, Alvaro Guarin, and Jose Norambuena-Contreras

Self-adaptive Construction Materials: Future Directions 215
Antonios Kanellopoulos, Magdalini Theodoridou, Michael Harbottle, Sergio Lourenco, and Jose Norambuena-Contreras

Fundamentals of Self-healing Construction Materials

Antonios Kanellopoulos and Jose Norambuena-Contreras

1 Introduction

The importance of materials science and engineering is paramount for developing new materials and components that will support our continuously growing needs as a society. At the same time, preserving our natural habitat by minimising our carbon emissions and waste of natural resources has also gained, quite rightly, significant attention over the last two decades. The concepts of structural integrity, durability and performance-based design have been the centre of attention of civil engineers over the previous two decades. In this challenging environment, self-healing materials have a significant role to play. Inspired by nature and the intrinsic healing process of living organisms, self-healing materials aim to tackle localised damage and restore the structural integrity of components.

Construction materials science is a fast-emerging field and a scientific topic that can lay the foundations of a sustainable future built environment. For many decades, the way we manufacture and use construction materials has not changed much, especially compared to other sectors. Construction materials have followed a very conservative evolution path. Nonetheless, in the last ten years, engineers have realised that the "future cities" can be realised by adopting the same standards, materials' properties, processing techniques and eventually carrying over the same problems and issues that materials used in construction have. Construction and building materials

A. Kanellopoulos (✉)
Division of Civil Engineering & Built Environment, School of Physics, Engineering & Computer Science, University of Hertfordshire, Hatfield, Hertfordshire, UK
e-mail: a.kanellopoulos@herts.ac.uk

J. Norambuena-Contreras
LabMAT, Department of Civil and Environmental Engineering, University of Bío-Bío, Concepción, Chile
e-mail: jnorambuena@ubiobio.cl

© Springer Nature Switzerland AG 2022

A. Kanellopoulos and J. Norambuena-Contreras (eds.), *Self-Healing Construction Materials*, Engineering Materials and Processes,
https://doi.org/10.1007/978-3-030-86880-2_1

are used to perform specific functions and satisfy a range of requirements stipulated by design codes and standards. While there are some advancements in aspects of construction materials used, for example, the complex geometries that we can construct today with concrete because of the rheological advances of the last twenty years; there is still a lot of work to be done to transform the functionalities and alleviate the limitations of traditional construction materials.

Ashby introduced the performance index concept where the performance of any material is a direct function of two components: its material properties and a shape efficiency factor which depends on loading conditions and the geometry of the individual components [1]. Construction materials face a continuous and somewhat unfair challenge. Everyone involved in the construction process (engineers, contractors, and clients) expect them to perform. The level of tolerance for lack of performance is limited. Simultaneously, design codes are based on mechanical functionality overlooking the fact that to ensure appropriate levels of mechanical performance the materials must maintain their overall integrity which in many cases is compromised by a combination of actions rather than mechanical fatigue. This need for change is crucial because it will address challenging technical issues such as early degradation. It will considerably improve the life cycle of infrastructure assets yielding a significant reduction in maintenance and replacement regimes. Eventually, this will lead to useless materials in the long run, and it will improve the users' confidence in the reliability and safety of structures. The focus of this book will be on the primary construction materials, providing an overview of strategies, assessment techniques, challenges and the outlook of self-healing applications. This book summarises the main studies recently developed on self-healing construction materials, emphasising in the fundamental concepts, monitoring and large-scale applications of these advanced materials. Due to its applied context, the contents of this book have been focused on the construction materials mostly used in the engineering field, such as concrete, asphalt, metals and alloys. The book should be attractive to the entire community of the science and engineering of materials, working on advanced construction materials. Finally, the technical level of the chapters allows it to be a guide textbook in undergraduate and graduate courses that have a focus on construction materials.

2 Historic Perspective

Natural resources have been the basis of growth in civil infrastructure for at least 100 years. At the same time, the exploitation of natural resources and the pollution associated with their extraction and processing have been already stretching the limits of our natural habitat, putting significant pressure on the environment and human well-being. According to the United Nations, the world's urban population has quadrupled since the 1950s, reaching 4 billion today, and with an estimated 7 billion people living in cities by 2050 [2]. To sustain this rapid urbanisation rate, billions of tonnes of construction materials are required annually. All evidence and projections suggest that civil infrastructure growth will keep expanding to meet the needs of modern urban societies.

The manufacture of construction materials is a process that utilises enormous amounts of raw materials and energy due to the sheer scale of the industrial processes required. Concrete is the dominant construction material and the critical element in most infrastructure assets. However, concrete manufacture on a gigatonne scale per annum imposes extremely high energy and resource demands: with >4 billion tonnes of cement being produced annually, accounting for ~8% of global anthropogenic CO_2 emissions, and yielding an annual production of ~2 tonnes of concrete for every person on the planet [3]. In Europe, the construction sector alone is responsible for 36% of CO_2 emissions and 40% of all energy consumption [4]. Cement manufacture is a significant contributor to the whole clinkerisation process consuming ~3.5 GJ/ton [5]. Of course, environmental burden and maintaining the structural integrity and serviceability are not unique to cement-based materials. The other two dominant players in construction, asphalt and steel, face similar issues. We need all these materials to support our growth. Still, at the same time, we cannot afford to waste natural resources, and most importantly, we cannot afford to produce infrastructure assets with long-term integrity problems.

As mentioned earlier, it is the coupled phenomena that lead to degradation, and this is accelerated with the formation of cracks either on the surface or the bulk volume of construction materials. Although ways have been developed to protect the materials (e.g. porosity reducing admixtures; coatings), the problem of cracking remains. The main issue with the crack repair is that it is not straightforward for several reasons. First, formed cracks may not be visible or may not have reached the surface. Secondly, there are quite a few instances where the cracks are not accessible. Besides the technical difficulties, there is also the cost and nuisance associated with the repair and maintenance of infrastructure assets.

Historically, the concept of self-healing in construction materials is not new. It has its roots back in the Roman era, where the lime-rich mortars used then showed remarkable durability over the millennia, predominantly due to the chemical interaction between the mortar constituents and the surrounding environment. This was not an intentional engineering intervention but rather a coincidental one. It was not until recently that scientists found this feature of the Roman mortars [6, 7]. There were some references about self-healing materials over the past century, but nothing was done systematically or materialised [8]. The first systematic attempt to develop self-healing processes in materials, quite coincidentally, started from concrete by Carolyn Dry in mid-1990s [9–12].

The breakthrough and step change in self-healing materials happened in 2001 with the landmark publication in Nature Journal by the White group [13]. In this work, the researchers have embedded microcapsules of lower stiffness to the host polymeric matrix. Once a crack propagated through the material and reached a microcapsule, this ruptured, releasing an adhesive that repaired the crack. This work caught the attention of materials scientists and engineers worldwide. It inspired several studies on this topic, not only in polymers/composites but also in construction materials. This

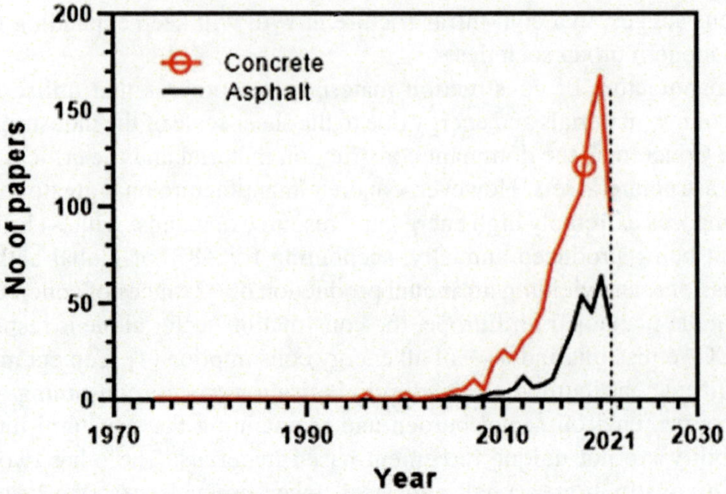

Fig. 1 Number of papers published on self-healing concrete and asphalt (*Source* Scopus; The data for 2021 concerns papers published the first quarter of 2021 only)

translated to many scientific papers published discussing self-healing principles and mechanisms in two highly used construction materials: concrete and asphalt (Fig. 1).

In asphalt, the earliest reported work on self-healing action dates back to 2003 [14], but it was not until the end of that decade when the topic lift-off [15–17]. In concrete, one can find scattered references on self-healing actions throughout the decades, as it is discussed in Chap. 2. Carolyn Dry's work was pioneering in the mid-1990s, but similar to asphalt, it was not until the end of the 2000s that systematic work on self-healing concrete was observed [18–21]. There is a strong incentive to develop a step change in the way materials are being designed and used in the construction sector. Materials' efficiency must be improved as it strongly associated with environmental impact through extensive maintenance and/or replacement regimes. Reinforced concrete structures are well known for their versatility, but they are equally known about their associated degradation problems. Likewise, highway infrastructure in the entire operation supports socio-economic growth, but disruptions, delays and public nuisance begin when issues arise. This is the reality with the two primary construction materials (concrete and asphalt) for many decades, but of course, steel structures and steel and alloys elements cannot be excluded from this discussion. Highlighting these issues does not serve the purpose of downgrading the significant progress over the last decades in manufacturing better construction materials. However, as engineers, we have to be pragmatic and recognise that construction materials are being improved. In the longer run, they still suffer the loss of structural reliability and deterioration and, as a consequence, reduce their durability.

3 Intrinsic and Engineered Healing

As material healing, we can define the recovery of some or all the properties of that material following damage initiation. The intensive research efforts of the last two decades led to the distillation of terms used to describe the mechanisms of healing in construction materials. Thus, self-healing can happen as a result of either of the following broad mechanisms (or a combination of both): *autogenous healing*, which relies on the intrinsic properties of the material and *engineered healing,* which a stimulated process by compounds and components that would not otherwise exist in a typical construction material matrix.

In cementitious systems, the former mechanism could be, for example, the continuous hydration of unhydrated cement particles. In contrast, the latter could refer to the release of an adhesive or mineral in the damaged area to promote repair. De Rooij et al. have in 2013 summarised all the autogenous healing mechanisms in cementitious matrices in three broad categories (see Fig. 2): physical, chemical and mechanical [22].

Of course, the exact manifestation of these mechanisms and how healing is promoted depends on the nature of each material. Continuous hydration works well for cementitious matrices, while elevated temperatures work well in metallic systems. In the subsequent chapters of this book, different materials and mechanisms will be considered and reviewed.

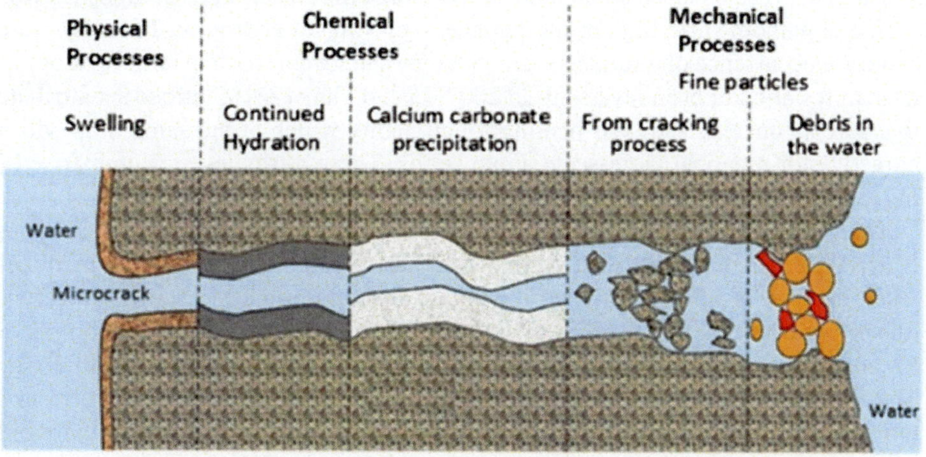

Fig. 2 Main mechanisms contributing to autogenous healing in cement-based materials (Reproduced from Rooij et al. [22])

4 Damage Prevention and Control

Dealing with damage effectively and for a considerable amount of time is the holy grail of materials science as it will enable increased functionality of components and higher levels of structural reliability. As mentioned earlier in this chapter, damage under service conditions is inevitable for any material. However, it should not be the damage leading actions upon the materials that define the performance of components. Instead, it should be our engineering/design interventions that allow materials and components to handle these actions. Therefore, there is a need to transform the way we design materials and elements.

With the progress of materials science and understanding how matter behaves, it became evident that crystal structures and molecular arrangements are essential elements that broadly define the response of materials to stimuli. With the development of advanced instrumentation, innovative manufacturing techniques and further understanding of micromechanics, scientists created a specific set of properties for the material. Advances in the field of fracture mechanics accurately predicted the formation of microdefects and their evolution, leading to the complete rupture. Information about defect creation and propagation was useful to redesign materials and components in such a way that damage is prevented. An early application of this principle was the use of steel reinforcement in concrete elements to enable the material to withstand flexural forces successfully. Nonetheless, the prime example was the rise of fibre-reinforced composites. Fibres incorporated in brittle and quasi-brittle matrices to prevent catastrophic failures. Although fibre-reinforced composites work perfectly without suffering abrupt damage, they still suffer from it. In a way, fibres avoid the occurrence of a sudden event, but they cannot control it in terms of stopping its manifestation or even reverse it. In 2007, Li and Yang [23] identified six attributes to create a concrete with self-healing functionality which at the same time will be attractive for practical large-scale applications. These attributes are summarised in Fig. 3.

Of course, these attributes can be extended to any construction material and any material in general. A self-healing material has to be able to respond immediately when damage occurs ("pervasive") while at the same time the self-healing functionality must remain active throughout the service life of a structural component ("stable"). Damage may occur at any stage of the lifecycle of an infrastructure asset. The construction sector is a bulk materials sector. The volumes used in civil infrastructure are staggering. Being financially viable for large-scale applications is extremely important for self-healing construction material ("economic").

In engineering terms, construction favours materials that have a predicted behaviour and perform reliably in various conditions. The expectation from a self-healing construction material is the same and is/will be expected to cover multiple performance and damage scenarios ("reliable"). Attaining good levels of healing fully recovering as many properties as possible is also very important ("high quality"). And last but not least, a self-healing material ideally must be able to counteract multiple damage events ("repeatable").

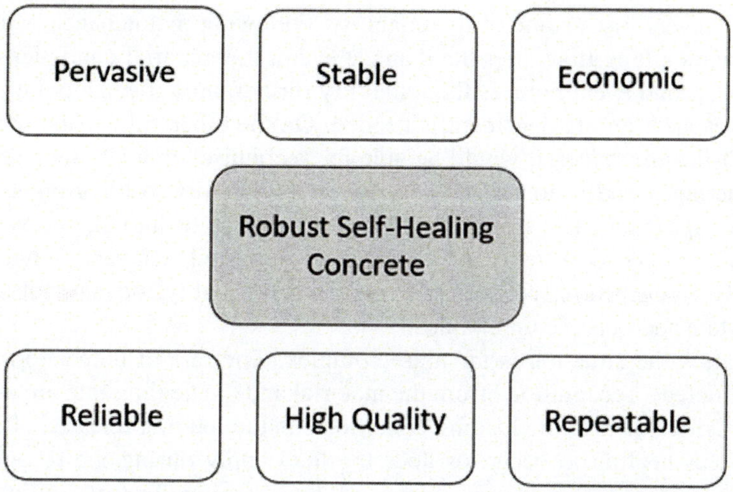

Fig. 3 Six attributes that need to be satisfied towards developing a robust self-healing concrete (according to Li and Yang [23]).

Self-healing materials aim to enhance materials' performance by controlling the effects of the damage on the performance of a component. Let us see a theoretical model of the reliability of a component concerning the service time and the occurrence of damage events. A non-self-healing material (see Fig. 4) exposed to a

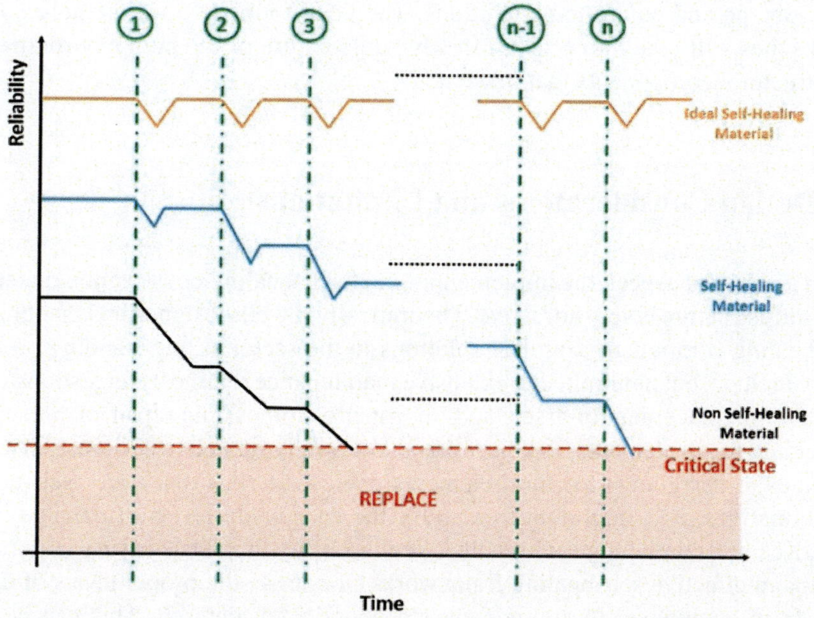

Fig. 4 Theoretical representation of the reliability of different types of materials concerning time and the occurrence of damage events.

"damage" event loses some of its reliability. Following maintenance, which sometimes occurs a long time after the damage event, the material continues to operate safely. A second event reduces the reliability further until at some point the critical state is reached, and replacement is needed. On the other side of the spectrum, an ideal self-healing material would be able to "recognise" quickly the damage event counteracting it and restoring the component's reliability to its original level after every damage event. The existing self-healing material technology provides a good recovery after the occurrence of damage, but repeated recovery is not achieved, especially in self-healing construction materials. Inevitably, the materials reach their critical state and become unreliable.

In reality, the situation is far more complex. The rate of defect formation, the speed of defect "recognition" from the material and the healing rate are all essential factors. Critical is also the loading conditions. Is the loading released after damage to allow for healing to occur, or does the load apply during the recognition and healing processes? As one can understand, these various factors can lead to large number of combinations and scenarios with varying complexity. Also, the healing processes may require minimum distances between crack-planes and even effective "communication" between "damage sensing" and "healing trigger" mechanisms. It is apparent that self-healing materials are not just "materials" but rather they are complex systems of individually engineered components.

Regarding construction self-healing systems, there is still a lot of work to achieve repeated performance and long-term reliability. Self-healing functionality in construction materials, and especially in asphalt and concrete, has been well documented over the last 15 years. Nonetheless, this functionality was demonstrated under very specific and controlled conditions. The need to move towards more complex approaches will take into account the dynamic nature of the built environment and the structural components that make it.

5 Design Considerations and Limitations

From a practical aspect, the implementation of self-healing construction materials in civil infrastructure is very attractive. Theoretically, a well-designed and manufactured self-healing component can offer solutions to the sector's long-standing problems. These include, but not limited to extensive maintenance regimes; congestions/delays; premature replacement of assets and improper or impossible repair/maintenance of difficult to reach elements. All these are also strongly related to sustainability. When it comes to designing structural elements using materials that have "self-healing" functionalities, it is important to identify the acceptable levels of damage in each case. Keeping components within their serviceability limit state is very important for having an effective self-healing framework. Of course, the proper levels of damage vary from component to component and material to material. This can be in the form of maximum acceptable crack as it is the case in concrete elements and asphalt materials, or it can also be in the form of maximum acceptable loss of strain recovery as it is the case in steel elements.

Another important factor to take under consideration is the source of damage. As mentioned earlier in this chapter, it is extremely rare to have a single source of damage on construction materials. It is the combination of mechanical and environmental weathering that will lead to the gradual manifestation of damage. With the process of time and repeated cycles of action, it can be detrimental to engineering components. To counteract this damage, it is important to understand the nature of the detrimental actions, which will help design the best course of action in deploying suitable self-healing mechanisms. What is certain is that there is no panacea in self-healing mechanisms, and not one approach can fit all.

Self-healing systems are multicomponent elements that, in most cases, trade-off their properties in exchange for long-lasting performance. Embedding microcapsules, for example, in a matrix to promote self-healing actions leads to a reduction in that matrix's mechanical performance. The medium becomes less homogenous, and preferential crack pathways are created. Similar issues arise with the use of vascular networks as well. It seems that we are downgrading the mechanical performance of our composites to achieve a better long-lasting performance that will depend less on external intervention.

All these create incompatibility with the current design codes, which in the future need to adapt and make provision for the large-scale implementation of self-healing materials. Practitioners and clients need a higher level of confidence from the design point of view regarding how the use of a "self-healing construction system" will affect the operation of the structure. Are there any knock-on effects from using such systems in bulk volumes and which are these? Currently, we do not have clear answers from the structural design point of view. Another practical barrier, at the moment, is the lack of standardisation in self-healing construction materials. The construction sector is very conservative in adopting new materials and technologies, and the lack of standardisation makes the application of self-healing materials difficult.

Although significant progress has been done the last twenty years, there is a lot of work yet to be done to create a solid framework that will yield robust self-healing construction materials. As researchers, we need to provide conclusive data and strong evidence that the materials we develop can perform: (i) outside the laboratory; (ii) at the large scale; (iii) under multiple and varying exposure conditions; (iv) repeatedly and (v) safely. In addition, we need to consider the dynamic nature of the civil infrastructure. We need materials that ideally can heal under service conditions, and they do that quickly and effectively. Shutting down a bridge, a tunnel or a highway to allow for extended healing periods is not much different from the existing situation. As people at the forefront of this scientific topic, we also need to provide enough information and investigate in detail the whole life aspect of our advanced self-healing materials. We aim for the "ideal" self-healing material with perpetual good performance, but the reality is that materials reach a point that needs replacement. How do our self-healing construction materials fit within the circular economy principles? In some cases, we do embed polymers, chemicals and even bacteria in them to promote healing. Will the material be able to be recycled in 150/200 years' time?

Addressing these technical considerations will make the way towards standardisation and implementation in the large much easier.

References

1. Ashby M (2010) Materials selection in mechanical design: Fourth edition. Butterworth-Heinemann
2. Oberle B, Bringezu S, Hatfield-dodds S et al (2019) Global resources outlook 2019: natural resources for the future we want. United Nations Environment Programme
3. Gale J, Bradshaw J, Chen Z et al (2005) Sources of CO2. IPCC Spec Rep Carbon Dioxide Capture Storage 77–103. https://doi.org/10.1021/es200619j
4. Ouldboukhitine S-E, Belarbi R, Jaffal I, Trabelsi A (2011) Assessment of green roof thermal behavior: a coupled heat and mass transfer model. Build Environ 46:2624–2631. https://doi.org/10.1016/j.buildenv.2011.06.021
5. Heravi G, Nafisi T, Mousavi R (2016) Evaluation of energy consumption during production and construction of concrete and steel frames of residential buildings. Energy Build 130:244–252. https://doi.org/10.1016/j.enbuild.2016.08.067
6. Jackson MD, Mulcahy SR, Chen H et al (2017) Phillipsite and Al-tobermorite mineral cements produced through low-temperature water-rock reactions in Roman marine concrete. Am Mineral 102:1435–1450. https://doi.org/10.2138/am-2017-5993CCBY
7. Witze A (2017) Rare mineral is the key to long-lasting ancient concrete. Nature. https://doi.org/10.1038/nature.2017.22231
8. van der Zwaag S (2007) Self healing materials: an alternative approach to 20 Centuries of materials science. Springer
9. Dry C (1994) Matrix cracking repair and filling using active and passive modes for smart timed release of chemicals from fibers into cement matrices. Smart Mater Struct 3:118–123. https://doi.org/10.1088/0964-1726/3/2/006
10. Dry C, McMillan W (1996) Three-part methylmethacrylate adhesive system as an internal delivery system for smart responsive concrete. Smart Mater Struct 5:297–300. https://doi.org/10.1088/0964-1726/5/3/007
11. Dry C (1996) Procedures developed for self-repair of polymer matrix composite materials. Compos Struct 35:263–269. https://doi.org/10.1016/0263-8223(96)00033-5
12. Dry C (2000) Three designs for the internal release of sealants, adhesives, and waterproofing chemicals into concrete to reduce permeability. Cem Concr Res 30:1969–1977. https://doi.org/10.1016/S0008-8846(00)00415-4
13. White SR, Sottos NR, Geubelle PH et al (2001) Autonomic healing of polymer composites. Nature 409:794–7. https://doi.org/10.1038/35057232
14. Saxegaard H (2003) Crack self-healing properties of asphalt concrete: laboratory simulation. Int J Hydropower Dams 10:106–109
15. Hao JT (2008) Important topics in development of asphalt concrete technology in China. Shuili Xuebao/J Hydraul Eng 39:1213–1219
16. Qiu J, van de Ven MFC, Molenaar AAA et al (2009) Investigating the self healing capability of bituminous binders. Road Mater Pavement Des 10:81–94. https://doi.org/10.1080/14680629.2009.9690237
17. Garcia A, Schlangen E, Van de Ven M (2010) Two ways of closing cracks on asphalt concrete pavements: Microcapsules and induction heating. In: Key Engineering Materials. Trans Tech Publications Ltd, pp 573–576
18. Jonkers HM (2007) Self healing concrete: a biological approach. In: Springer series in materials science. Springer Verlag, pp 195–204
19. He H, Guo Z, Stroeven P et al (2007) Self-healing capacity of concrete-computer simulation study of unhydrated cement structure. Image Anal Stereol 26:137–143. https://doi.org/10.5566/ias.v26.p137-143

20. Granger S, Loukili A, Pijaudier-Cabot G, Chanvillard G (2007) Experimental characterization of the self-healing of cracks in an ultra high performance cementitious material: mechanical tests and acoustic emission analysis. Cem Concr Res 37:519–527. https://doi.org/10.1016/j.cemconres.2006.12.005
21. Li V (2009) Self-healing concrete repairs cracks, maintains strength. Mater Perform 48:20–22
22. de Rooij MR, Van Tittelboom K , De Belie N, Schlangen E (2013) Self-healing phenomena in cement-based materials: State-of-the-art report of RILEM Technical Committee 221-SHC: self-healing phenomena in cement-based materials. Springer
23. Li VC, Yang EH (2007) Self healing in concrete materials. In: Springer series in materials science. Springer Verlag, pp 161–193

Self-Healing Cement-Based Materials: Mechanisms and Assessment

Antonios Kanellopoulos

1 Introduction and Historic Perspective

Reinforced concrete (RC) is the dominant construction material and the key element in the vast majority of infrastructure assets, which in turn are the backbone of our societal and economic growth. Bridges, highways, air- and seaports, powerplants, water and sewage treatment facilities, hospitals, tunnels are all essential to maintain the modern way of life. However, concrete's manufacture is extremely energy and resource intensive: > 4 billion tonnes of cement produced annually, accounting to ~8% of global anthropogenic CO_2 and resulting to an annual production of ~2 tonnes of concrete for every person on the planet [1]. Despite the huge environmental impact, concrete and the related cement-based materials are the construction industry's favourites for a variety of reasons: (i) it is a global material, readily available almost everywhere; (ii) ease and cost of construction compared to alternatives (e.g. steel); (iii) robustness for a variety of exposure scenarios; (iv) ability to construct a large variety of complex geometries and (v) excellent mechanical performance.

The exposure of cement-based infrastructure to a large number of degrading environments throughout their service life cannot be prevented. The integrity of concrete and cement-related materials largely depends on their ability to withstand mechanical and environmental weathering. It is the coupled effect of phenomena such as impact, service loading, chloride/CO_2 concentration, pressure/thermal differentials, freeze–thaw and sulphates that can cause severe damage, hence reducing functionality. Damage can manifest itself by localised weakening of the material and can progress in the form of microcracks that gradually reduce the integrity of structures and elements. Under continuous mechanical and environmental stress, these

A. Kanellopoulos (✉)
School of Physics, Engineering & Computer Science, Division of Civil Engineering & Built Environment, University of Hertfordshire, Hatfield, Hertfordshire, UK
e-mail: a.kanellopoulos@herts.ac.uk

© Springer Nature Switzerland AG 2022

A. Kanellopoulos and J. Norambuena-Contreras (eds.), *Self-Healing Construction Materials*, Engineering Materials and Processes,
https://doi.org/10.1007/978-3-030-86880-2_2

microdefects coalesce and expand compromising the mechanical properties of materials. In addition, the microcracks lead to the rapid rise of corrosion of the steel reinforcement by creating paths for oxidising species. Corrosion-related costs are a major problem worldwide with a global annual budget of 2.2 trillion US dollars. In the UK alone, the annual cost of corrosion comes to a total of £70 billion [2] where currently between 35 and 40% [3] of all construction spending is on repair and maintenance of civil infrastructure.

The design codes, such as BS EN 1992 and ACI 318, consider the degradation of concrete as inevitable and as such they prescribe a very strict set of rules and factors. Although cracking cannot be totally prevented, in fact design codes permit and allow cracking levels up to a certain extent, degradation of materials and components can. In concrete and cement-based materials, it is the structure and the connectivity of the pore network that will define the permeation properties of the material. These properties affect the way water soluble and/or air-borne deleterious species can permeate and/or diffuse into the material. Chlorides, carbon dioxide and sulphates are the most important species which, when they penetrate the material, gradually lead to material weakening, crack formation and eventually to the oxidation and corrosion of the steel reinforcement. Of course, chemical resistance of the cementitious matrix itself is also a very important factor for the longevity and integrity of elements.

The history of scientific investigation on self-healing cement-based materials goes back to 1836 where the French Academy of Sciences first documented as "healing phenomenon" the carbonation of portlandite produced during the hydration reactions [4]. It was later the pioneer concrete technologist, Duff Abrams, who observed that cracks forming during pull-out tests on reinforced concrete, "healed" when specimens were left to rest for a period of time post-testing. Abrams was convinced that the observed healing was "the effect of retarded or interrupted hydraulicity of cement" [5]. It was not for another 43 years until the first documented comprehensive investigation on healing of cementitious matrices [6]. In this pioneering piece of work, Lauer and Slate not only scientifically confirmed what the French documented more than a century before but they also proved that the actual healing on cementitious matrices was due to proliferation of both portlandite and calcium carbonate crystals.

In 1974, Wagner identified the autogenous (intrinsic) healing of concrete in mortar linings of water pipes [7]. In the same year, Ivanov and Polyakov published their study on durability improvement of cracked concrete as a result of healing processes [8]. As mentioned in Chap. 1, it was Carolyn Dry's work that had all the elements of a systematic approach on studying self-healing processes in concrete. Between the published works in 1974 and 1994 (Dry's first known published paper), there were some scattered papers but still no significant volume of work [9–12].

From early 2000s, more intense and systematic studies started appearing in the literature with increasing number of groups working on this topic, especially from the end of 2000s onwards. Early detailed research focused on intrinsic healing phenomena in typical concrete matrices [13–15]. Later research started expanding on biological assisted healing in cementitious matrices [16, 17], healing of fibre-reinforced cement-based materials [18, 19], the effect of supplementary cementitious materials on healing [20, 21], the influence of smart additives on promoting healing

[22], development of microencapsulated systems [23, 24] and vascular networks [25].

Since 2011, the area has seen great growth in terms of the volume of published work which in turn gradually shifted in more complex areas. Crystalline admixtures and other mineral additives were studied [26–28], the potential of shape-memory polymers to promote healing was studied [29, 30], biological healing further explored [31–33] and new encapsulation methods were developed [34–36]. This growing interest was also confirmed by the funding provided by national and international funding organisations. In the UK, the leading consortium in self-healing construction materials research is the Resilient Materials for Life which started in 2013 [37] and has given the first large-scale trial of self-healing concrete mechanisms in the UK [38]. Netherlands and Belgium have also been at the forefront of self-healing exploration with groups at TU Delft and Ghent University producing a large amount of significant scientific outputs. Netherlands also pioneered the large-scale application of this technology in asphalt pavements initially but also with concrete structures recently given the opportunities to new start-ups to emerge [39].

2 Healing Mechanisms in Cement-Based Materials

Healing mechanisms in cement-based materials have drawn significant attention the last 15 years. There are two distinct healing strategies: autogenous (intrinsic and stimulated) and engineered. The former takes advantage of the mineral chemical processes in a cementitious matrix and the latter involves the incorporation of components specifically designed to promote healing. Although self-healing phenomena in cementitious matrices are known for almost a century, it was not until the last two decades when we started decoding carefully these processes and gaining fundamental insight of the mechanisms.

2.1 Autogenous Healing (Intrinsic)

There is a number of published works that thoroughly revise the intrinsic healing processes in cement-based materials [40–44]. As discusses earlier (Fig. 1.2), intrinsic healing can be due to (or a combination of) complicated chemical, physical and mechanical processes. Here we will mainly focus and the physicochemical mechanisms rather than the blockage of cracks due debris. The cementitious matrices owe their intrinsic healing properties in their mineralogy. There is a large number of studies in the literature, and while there is significant scatter in the obtained results, they all seem to agree that delayed hydration of previously unhydrated cement particles and, more importantly, the carbonation of portlandite are two major reasons contributing to the manifestation of healing in cement-based matrices.

In the case of previously unhydrated cement particles, the process is straightforward. Once a crack is formed, it allows moisture to penetrate the material and once the moisture front comes into contact with the cement particles and the series of hydration reactions are initiated:

$$2C_3S + 6H \rightarrow C_3S_2H_3 + 3CH \tag{1}$$

$$2C_2S + 4H \rightarrow C_3S_2H_3 + CH \tag{2}$$

$$C_3A + 3C\hat{S}H_2 + 26H \rightarrow C_6A\hat{S}3H_{32} \tag{3}$$

$$2C_3A + C_6A\hat{S}_3H_{32} + 4H \rightarrow 3C_4A\hat{S}H_{12} \tag{4}$$

$$C_4AF + 10H + 2CH \rightarrow C_6AFH_{12} \tag{5}$$

The produced calcium-silicate-hydrate ($C_3S_2H_3$) and calcium hydroxide (CH) occupy more volume than the original cement particle they came from. As they form, they expand in the space created during the crack formation and eventually these hydration products heal the crack. It is apparent that this mechanism requires unhydrated cement particles in the system and the existence of such particles depends on a number of factors. On the other hand, the carbonation phenomena and crystallisation of portlandite are thought to be the primary mechanisms of intrinsic healing in cementitious matrices. Here we have two separate processes. First the carbonation of portlandite which is known as hydration product that dissolves in water. Once a crack forms, portlandite reaches the crack faces and in presence of moisture has the potential to homogeneous phase carbonation according to following reaction.

$$Ca(OH)_2 + CO_2 \rightarrow CaCO_3 + H_2O \tag{6}$$

Similarly, free calcium ions from the cement hydration reaction when exposed to moisture and CO_2 lead to the formation of calcium carbonates. The chemical sequence of this process is as follows:

$$H_2O + CO_2 \leftrightarrow H_2CO_3 \leftrightarrow H^+ + HCO_3^- \leftrightarrow 2H^+ + CO_3^{2-} \tag{7}$$

$$Ca^{2+} + CO_3^{2-} \leftrightarrow CaCO_3(pH_{H_2O} > 8) \tag{8}$$

$$Ca^{2+} + HCO_3^- \leftrightarrow CaCO_3 + H^+(7.5 < pH_{H_2O} < 8) \tag{9}$$

Edvardsen has shown that the carbonation of calcium ions in the vicinity of the cracks can initiate even at non-alkaline water pH values [13]. In fact, in some cases carbonation may start even in slightly acidic conditions. As the moisture front enters

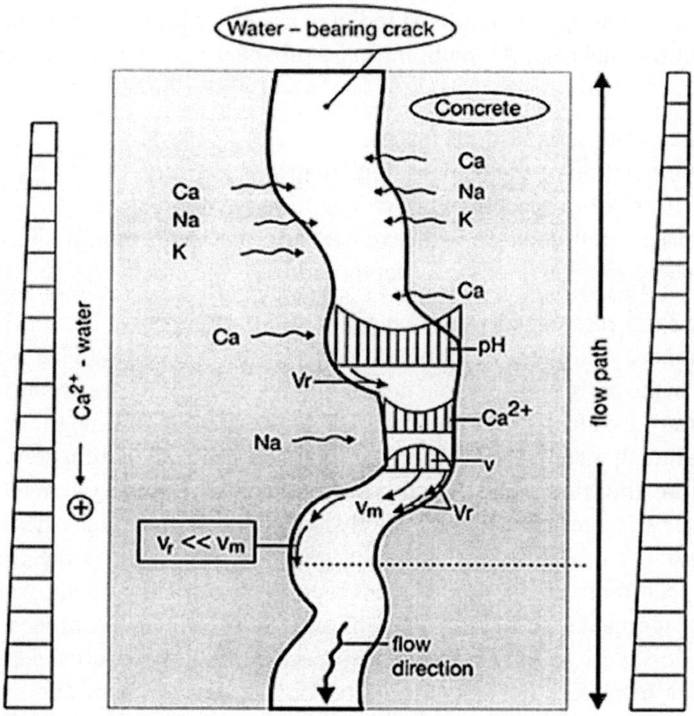

Fig. 1 Conditions in water-penetrated concrete crack [13]. Image reproduced with permission from ACI

through the crack, the highly alkaline environment of a cement-based material, it will locally shift towards the alkaline spectrum. At the same time, the cementitious matrix will besides calcium ions will also release sodium and potassium ions creating the appropriate conditions for precipitation of carbonation products. Figure 1 shows the conditions in water-penetrated concrete as described in Edvardsen's model.

It is obvious from the above discussion that in both cases, the sources that lead to intrinsic self-healing are not infinite. Unhydrated cement grains eventually hydrate and in the carbonation process, after the initial crystal growth stage the cementitious matrix in the vicinity of the crack becomes less rich in the elements promoting precipitation of carbonates. Besides the general agreement of the published studies about the mechanics of intrinsic healing, there is disagreement on the levels of healing attained, both in terms of size of crack healed and properties restored. In most cases, this type of healing can restore cracks up to 100 μm. Healing of cracks up to 300 μm is reported in some cases, but this necessitates a combination of parameters (e.g. high cement content, low w/c, presence of water). There is also significant chance that the phenomena of continuous hydration and carbonation take place at the same time, hence making the differentiation of the individual contribution a rather difficult task.

Intrinsic self-healing is difficult to predict or assume without having enough information about the material. As such, intrinsic self-healing depends on the following factors:

i. The age of the cementitious matrix.
ii. The composition of the cementitious matrix.
iii. The existence of supplementary cementitious materials.
iv. The presence of moisture, either in the form of vapour or in liquid form.
v. The geometric characteristics (depth/width) of the crack.

From the early referenced works on self-healing, it is known that intrinsic repair of cracks is more dominant in early age concrete [6, 7, 10, 13]. This is somewhat expected as in the early age concrete not all of the cement particles are fully hydrated and hence, they are available for reaction. With the process of time though it is also well-known fact that concrete hydration continuous and carbonation comes also into play. Therefore, the "raw" components, i.e. the calcium ions supply, that feed intrinsic healing at the early ages are either depleted or available at small quantities that cannot offer a significant healing effect. Intrinsic healing is affected by the cementitious matrix composition in all aspects: cement type used; cement content; water content and type of aggregates. The way cement and water affect this process has been discussed extensively so far. On the aggregates side, their type (crushed/uncrushed) and their size have significant contribution on the development and evolution of cracks influencing their geometric characteristics. In addition, there are cases where aggregates can span across a crack plane influencing the proliferation of healing products. The existence of supplementary cementitious materials (SCMs) in the system can be catalytic in the development of intrinsic healing processes. In the presence of suitable conditions (Ca^{2+} ions and moisture), SCMs can promote the formation of a surplus of hydration products that in turn can heal formed cracks.

It is evident so far that the presence of water is extremely significant parameter in the evolution of intrinsic healing mechanisms. The chemical reactions discussed earlier require the presence of water. Nonetheless, although water present in any form is certainly better than no water, there is now a general consensus that even exposure to high humidity cannot be enough to trigger such mechanisms to full extension and a complete immersion is required to promote healing [35, 44]. The effect of moisture was also proven in the first UK large-scale application of self-healing systems where the ambient moisture conditions appeared to have an impact on healing processes as seasons changed [38]. This of course raises some concerns on the efficiency of intrinsic healing in practice where complete immersion is not always possible or practically feasible or where there is not enough moisture/water vapour.

When it comes to the geometric characteristics of cracks, it is obvious that the smaller the crack opening and the shallower the crack depth the easier the crack will heal. Intrinsic self-healing is anyway very sensitive on the crack size, but the geometry of the cracks is also a significant parameter in other healing mechanisms as well. Controlling the crack size is a way that can be very effective in promoting healing mechanisms. Published research shows that the bridging action of fibres in fibre reinforced cementitious matrices has a significant, positive, effect in assisting the manifestation of intrinsic healing [18, 40, 41, 45, 46].

2.2 Autogenous Healing (Stimulated)

It is very rare to find today concretes without any mineral additions. First, most concretes are produced with blended cements that incorporate SCMs. Furthermore, SCMs and other mineral additives are also used widely in addition to blended cements for improving the performance of the produced cementitious matrix [47]. Stimulated autogenous healing has a similar mechanism to intrinsic autogenous healing. It is based on the principle that a considerable amount of the mineral additions will not be consumed in chemical reactions and will remain dormant until the necessary conditions rise that will promote their reaction. Since the principle is the same, it is apparent that certain restrictions are also common. For example, the necessity for high levels of moisture/water supply into the crack for healing processes to initiate and evolve.

Siliceous and alumino-siliceous additions (e.g. silica fume; fly ash; ground granulated blast furnace slag) have been widely used in the production of engineered cementitious composites (ECCs) and other high-performance matrices. It is well documented fact that the pozzolanic reactions of such additions result in increased amounts of calcium silicate hydrates (CSH) and the reduction of portlandite amounts [48]. It is also a known fact that such additions are either not fully consumed in hydration reactions at an early stage or they react at slow rates [49]. Therefore, they can remain dormant into the cementitious system for a long period of time. For fly-ash-rich matrices, it was found that pozzolanic reactions leading to healing can take place even up to one year post-casting [50, 51]. In the case of a crack formation and the existence of high levels of moisture, these siliceous/alumino-siliceous additions have the potential to engage into chemical reactions (pozzolanic activity) and form products that can heal the crack. There is evidence that low water-to-binder ratio matrices benefit more in terms of self-healing actions, predominantly because there is significant amount of unreacted mineral source to promote such actions [52–54]. The presence of calcium ions is a critical factor in the evolution of healing processes in SCMs containing systems. In low portlandite concentrations for example, slag was found to perform better compared to fly ash [53, 54]. It seems that the higher CaO concentrations found in slag provide the necessary calcium ions that can promote the proliferation of calcite crystals [55].

Over the years, research has also focused on other mineral additions in relation to healing. Minerals such as montmorillonite, quick lime and magnesium oxide have been investigated. Montmorillonite, in the form of bentonite clay, is an assembly of a 2:1 gibbsite sheet between silica sheets and has significant swelling properties than can reach 15 to 18 times its dry weight. In highly alkaline environments, montmorillonite follows a dissolution process that precipitates different crystals that range from zeolites, phillipsites, CSH, saponite, etc. depending on the exact environmental conditions (pH & temperature ranges). These precipitates were found effective to facilitate self-healing in cement-based systems [56, 57]. Similarly, quick lime (which is almost pure CaO) is also known for being highly expansive in contact with water, a characteristic which is also found in the hydration processes of reactive magnesium oxide [58]. Recent research showed that there are optimum combinations of these expansive minerals which can significantly improve the autogenous healing capacity of cement-based matrices. Qureshi et. al. have shown that ternary blends of bentonite, lime and MgO have improved healing performance both at early and later stages of the life of the cementitious matrix [59, 60]. Healing products found to form in the perimeter of induced cracks (see Fig. 2). These products reported to be dense and highly crystalline. Detailed microstructural analysis has shown that the addition of these expansive minerals promoted the precipitation of brucite, other magnesium hydrocarbonate products and combined magnesium-calcium and alumina hydration products.

Since 2015, there is also a growing interest for the so-called crystalline admixtures (CA) and their role in assisting self-healing in concrete. Crystalline admixtures are not considerably different on molecular or crystalline level from other mineral admixtures. The term CA mainly reflects a commercial range of products which are used to enhance concrete properties and predominantly are used as permeability reducer admixtures (PRA) [27]. Crystalline admixtures are made of chemicals mixed, some of which are labelled as "crystalline promoters", with cement and sand. CAs are hydrophilic, and upon contact with water, the crystalline promoter interacts with the alite (C_3S) in cement to produce a modified form of CSH. Research to date has demonstrated the effectiveness of such admixtures in a variety of scenarios and combinations [27, 61–63]. Although CAs are commercial products with market that have proven self-healing effect, still there are many elements that need to be clarified.

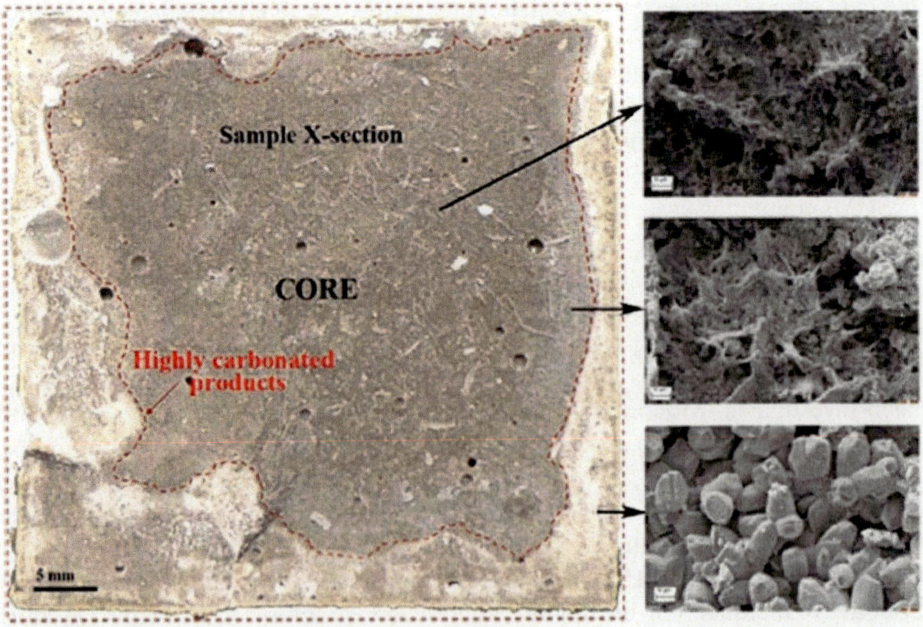

Fig. 2 Healing material formation on the perimeter of an initially damaged prism (left) and the corresponding SEM images from specific locations of the crack plane. Image reproduced from [59]

There is a wide variety of different types of CAs and just because they are branded products proprietary issues have not allowed yet deeper investigation on the exact mechanisms of healing evolution. In addition, it is the opinion of the author that the wide variety of CAs, as far as the self-healing effect is concerned, does not provide adequate levels of reassurance needed by the end user.

2.3 Engineered Healing

Engineered healing depends on the incorporation within the cementitious matrix of materials/compounds/systems that are not typically present in concrete. Following from the early attempts to promote healing in concrete through macroencapsulation [64] and the breakthrough advancements in microencapsulated and vascular self-healing in polymers [65, 66], these techniques have drawn a lot of attention in the area of construction materials. Encapsulation, in various forms and arrangements, has been a favourable method to promote healing mechanisms in concrete and asphalt. Over the years, engineers and scientists have found value in using other unconventional compounds to enhance the self-healing capabilities of cementitious matrices. These include the use of: (i) various bacterial strains; (ii) polymeric compounds with tuneable properties; (iii) mineral nano-fillers and (iv) other nano-compounds. From the engineering perspective, it is obvious that the addition of such "alien" compounds into the cementitious matrix bares a number of technical challenges that include, but not limited to, issues around hydration kinetics and setting times, mechanical performance of the composite, attenuation of adequate healing levels, survivability of embedded compounds, etc. Figure 3 summarises the techniques used in engineered healing processes.

Developing self-healing mechanism using encapsulation is based on the simple concept that a healant compound remains dormant within a protective shell which

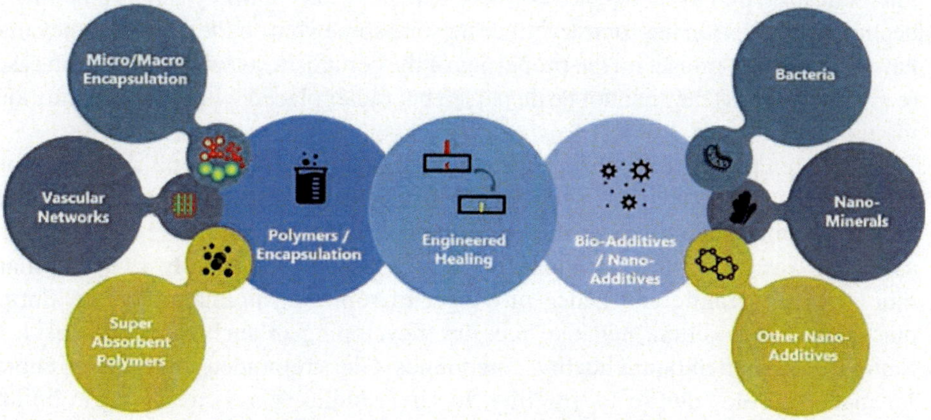

Fig. 3 Techniques used to promote engineered healing in cementitious matrices

ruptures when it meets with a propagating crack front. Rupture of the shell will result in the release of the healant in the crack plane, and through a series of chemical reactions healing will occur. In the landmark work by Scott's group, microencapsulation promoted self-healing in polymers utilised a two-part adhesive [65]. The resin was encapsulated while the catalyst was mixed as part of the matrix. Rupture of the microcapsule brought resin and catalyst together activating the crosslinking sequence of the adhesive, a sequence which eventually led to the healing of the crack. Albeit this method proven to be very effective in thin profile polymeric laminates when it came to bulk materials like concrete, the encapsulation and more specifically microencapsulation of adhesives and resins revealed several technical challenges. First, the use of two-part adhesives was not a feasible option as the amount of catalyst needed to be dispersed in the cementitious matrix was substantial. Secondly, the use of single part adhesives (e.g. cyanoacrylates) showed either low degree of penetration into the crack plane or premature polymerisation even before the adhesive leaves the protective shell [67, 68]. While the benefit of quick reaction of adhesive compounds is not disputed, the effectiveness of encapsulated epoxies is highly questionable. The long-term stability and functionality of epoxies is at question. In addition, all epoxies are strongly linked with aggressive chemicals which if used as additives in a bulk material like concrete can lead to significant sustainability issues in the longer run.

2.3.1 Macroencapsulation and Vascular Systems

Macroencapsulation was one of the first areas that attracted interest, predominantly as a consequence of Dry's work in mid-1990s where glass capillary tubes were used as carriers for the healing compounds [64, 69]. With the term "microcapsule" is defined any encapsulated system that is larger than 1 mm in diameter or in length. It can either be in the form of hollow tubes filled with healing compounds or in the form of pellets either made from healing compounds or impregnated by them. The benefits of the macrocapsules are that can be manufactured and assembled easier than the microencapsulated systems and that they can carry substantially larger amounts of healing materials. On the contrary, since they are somewhat "bulky" carriers they may have a significant impact on the properties of the composite as well as in certain cases (e.g. hollow tubes) they cannot be dispersed but rather placed within the cementitious matrix.

Early attempts on the macroencapsulation front utilised small capillary glass tubes to either secure the adhesives in them or to create a micropipe continuum where adhesives will be continuously fed into the crack [15, 70, 71]. The capillary tube approach was not further developed by other researchers, mainly since capillary tubes are very fragile and hence in danger of rupturing prematurely (e.g. during placement). In addition, negative pressure developing in such tubes can result in limited dispersion of liquid healing compounds. Glass reminded a material of choice for macrocapsules due to its fragility. In later studies, glass capsules and hollow tubes were used to carry a variety of healing agents. Some studies encapsulated polymeric precursors [72, 73], whereas others encapsulated liquid minerals [35].

Polymeric precursors interact with moisture through a chemical reaction that leads to the production of expansive products that fill the formed cracks. Liquid minerals on the other hand engage into mineralogical reactions like the ones described in the autogenous healing section resulting in the proliferation of crystals that heal the formed cracks. As with autogenous healing, the existence of significant amounts of water was found to be very critical in both cases. Qureshi et. al. used the glass capsules to carry powdered expansive minerals, and to solve the problem of dispersion, they devised a concentric system of glass capsules where the inner capsule was carrying water and the outer capsule the powder minerals [74].

Nonetheless, the glass macrocapsules may contribute to the development of alkali–silica reactions which are known to be detrimental for cementitious matrices. To overcome this issue, some researchers recommended the use of extruded macrocapsules made either from a cementitious matrix or from polymers (e.g. poly(lactic acid); polystyrene and poly(methyl methacrylate/n-butyl methacrylate) [75, 76]. Both types showed good healing performance; however, an analytical investigation of the extruded polymeric capsules showed that their effectiveness was strongly correlated with their thickness. More specifically, it was shown that best performance was exhibited by polymeric capsules with wall thickness between 0.3 and 0.7 mm, as larger wall thickness would require a substantial crack opening for capsule fracture to occur [77]. Others have shifted away from tubular type of macrocapsules and investigated the potential of using impregnated lightweight aggregates (LWA) and mineral pellets to promote self-healing. Sisomphon et al. used sodium monofluorophosphate (Na_2FPO_3) solution to impregnate LWAs under vacuum [78]. The impregnated LWAs were incorporated in slag cement mortars, and the results showed that they contributed in the healing of cracked sections. Alghamri et al. followed a similar approach impregnating LWAs under vacuum with sodium silicate [79]. Nonetheless, in this case a thin polyvinyl alcohol coating was applied to the impregnated LWAs to contain and control the release of sodium silicate. The results of this study showed successful release of sodium silicate and the formation of calcium silicate hydrates and carbonation products in the crack planes. Recently, Wang et al. used cement-based healing pellets (CHP) to enhance healing actions in cementitious matrices [80]. CHPs were made with cement and sodium carbonate which was used to boost the carbonation processes. In a similar fashion to other mineral-based approaches, this study demonstrated healing of cracks as a result of extensive carbonation activity in crack.

The use of tubular macrocapsules is extensive in the so-called ***vascular healing*** approach. As the word signals, the vascular approach aims to mimic the circulatory system found in living organisms. A network of artificial veins is created and embedded in the cementitious matrix. Upon crack formation, the network ruptures in the location of the crack and healing compound is run through the system and fills the crack plane. The early attempts in vascular healing involved one-dimensional networks, in essence a single hollow tube that was fed externally with healing material [67]. With the process of research, the development of vascular networks became more complex and expanded into 2D and 3D networks [81–83]. Compared to other healing systems, vascular networks have the benefit that can provide a continuous feeding of healing material to the crack. In addition, if the vascular network is properly pressurised, the issue of improper healant distribution in crack plane can be resolved. Published studies so far have examined the use of both epoxies and liquid minerals as healing materials. This area is currently under development as many parameters need to be clarified to yield optimum healing. For this reason, recent studies focus on the flow characteristics and healing mechanisms of this approach [84, 85].

2.3.2 Microencapsulation

Looking into the literature, it is evident that ***microencapsulation*** is covered in most "encapsulation"-related studies on self-healing concrete. Scientifically is very attractive topic and technically has few advantages: (i) wider distribution within the matrix than macrocapsules; (ii) less disruption of the matrix homogeneity and (iii) in many cases microcapsules can be triggered by very small cracks addressing the problem quite early. The functionality of the microcapsules in self-healing cement-based matrices has been investigated in a number of studies the last decade [24, 86–89]. The shell and cargo materials vary as well as the production methods. Most of the studies do use emulsion polymerisation techniques to yield microcapsules that have poly-urea and/or urea–formaldehyde as shell materials and mostly epoxies as healants. The major issue of these systems is their compatibility with the host matrix as well as potential health and safety implications. The chemicals used in these processes are quite aggressive in most cases and the safety of bulk inclusion of such microcapsules in concrete is questionable. For this reason, some researchers turned into more environmentally friendly compounds such as sodium alginate, gum Arabic and gelatine [32, 36]. The core principle of emulsion polymerisation techniques is that two immiscible liquids are mixed together under agitation and controlled conditions (e.g. temperature and pH). Due to the immiscibility, droplets are formed in through a series of chemical processes, which are outside the scope of this chapter, a shell wall is formed on the perimeter of the oil droplets, hence encapsulating them (Fig. 4). The polymerisation process that yields the shell of capsule can be the

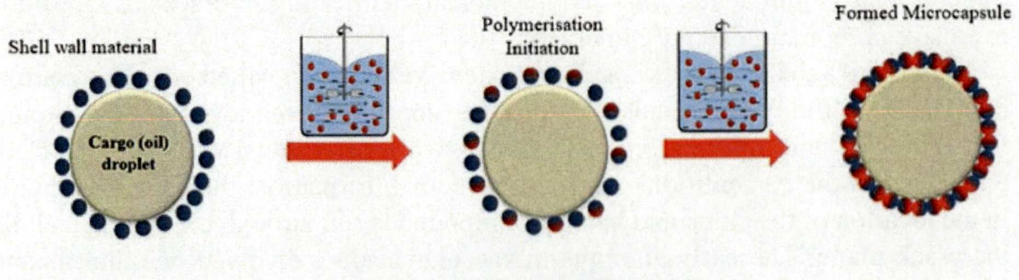

Fig. 4 Schematic representation of the encapsulation process in emulsion polymerisation

result of purely chemical crosslinking (interfacial polymerisation), the result of ionic attraction (complex coacervation) or the result of ion exchange (ionic gelation).

A technical challenge with such encapsulation production methods is that they are very sensitive to the relevant parameters (e.g. temperature) needed to promote crosslinking. In addition, under agitation is highly unlikely to produce monodisperse particles. There is always a range of sizes produced, and the best you can achieve is with very good fine tuning is to reduce the particle size distribution. In addition, controlling the thickness of the shell wall is also a challenge. Recent advancements in the field use microfluidics to encapsulate healants. Microfluidics is the area that studies the manipulation of fluids within channels that have diameters in the fraction of a micrometre. It is very popular in chemistry, biology and pharmaceutics. Souza et al. studied microfluidics as an alternative solution to microcapsule fabrication [90, 91]. A double emulsion was created in the microfluidic system, and monodisperse droplets were formed which in turn photopolymerised with ultraviolet light (Fig. 5). The outcomes of this work showed that the fabrication of one-size microcapsules with fully tuneable shell thickness is possible. The formed particles found to have good retention of the cargo. They appeared to survive when embedded in a cementitious matrix and fracture when a crack formed. One limitation of the microfluidic approach is its scalability. To produce large volumes of microcapsules, state-of-the-art systems are required, and this can have a knock-on effect on cost. Nonetheless, to be fair, this is a considerable challenge for all microencapsulation techniques. As mentioned earlier, the technical challenges and the volumes required for construction make this area a very interesting topic for research and development in the years to come.

In terms of cargo materials in microencapsulated systems, although some early attempts used epoxies, most of the studies now explore the possibility of minerals in liquid form. The level of healing, the rate of recovery and the exact processes depend on the exact system used each time. One can find a wide variety of microencapsulation systems in the literature. Where minerals are used the process of healing has the same

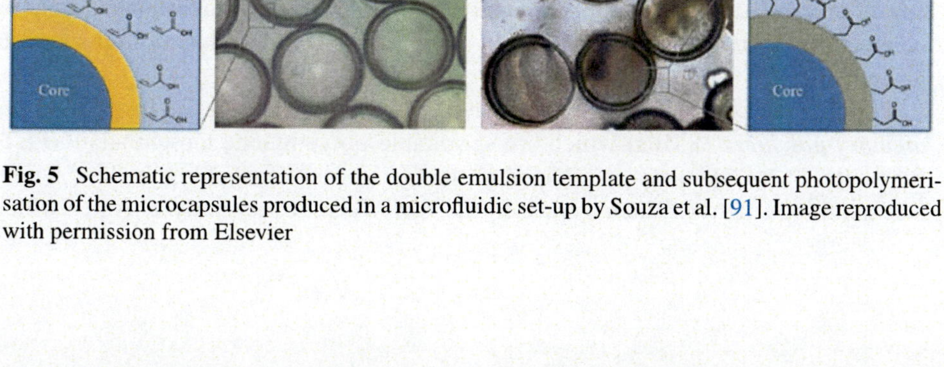

Fig. 5 Schematic representation of the double emulsion template and subsequent photopolymerisation of the microcapsules produced in a microfluidic set-up by Souza et al. [91]. Image reproduced with permission from Elsevier

concept as other mineral promoted healing discussed in this chapter. Nonetheless, still there is no universal consensus for a specific type of healing agent. This is valid not only for the microencapsulated systems but also for all encapsulated platforms. Another significant parameter that has not been yet widely investigated is the effect of microcapsules' type and dosage on the properties of the cementitious composite. Although there are some studies reporting a reduction on mechanical properties with incorporation of microcapsules, there is no a wide systematic study on topic [34, 36, 92]. A possible reason for this is that researchers use different systems and different types of microcapsules, which makes a round-robin test challenging.

2.3.3 Super Absorbent and Shape Memory Polymers

Staying on the polymeric side of things, a compound that has drawn a lot of attention in concrete self-healing research and development are the ***super absorbent polymers*** (SAPs). SAPs can be either natural or synthetic and are crosslinked polymers that can absorb liquid volumes that are considerably larger than their original volume. SAPs retain the liquid and slowly release it depending on the surrounding conditions. They are widely used in agriculture and sanitation products industries and the recent years have attracted the attention of construction for a variety of applications which include control of rheology, shrinkage and adverse durability effects [93]. For self-healing applications, SAPs in the form of miniature beads are mixed with the cementitious matrix. The polymers remain dormant in the matrix, and when a crack forms, the moisture that will penetrate the crack will cause their swelling and hence promoting the blockage of the crack plane. The concept has been demonstrated very well from a number of researchers [22, 46, 94, 95]. The mechanism of healing is similar to what has been discussed before. SAPs act as local water reservoirs which promote hydration of adjacent unhydrated cement particles as well as the formation of carbonate products. To enhance mineral induced healing some researchers alongside with SAPs quite successfully have been using mineral doping of cementitious matrices with crystalline admixtures, fly ash and nano-silica [95–97]. Technically, the major issue with SAPs is that the presence of moisture is vital for them to function. In addition, similarly to microcapsules they do modify and affect the properties of matrix. At the end of the day, they are inclusions. Contrary to microcapsules, SAPs when they release the water, in other words post-mixing/setting, they lose considerable amount of their volume. This creates voids in the matrix, which will easily attract cracks. One can argue that this beneficial as the crack path and location can be predicted, and it would be a valid point. On the other hand, though, creating a cementitious matrix with significant volume of voids will have an impact on its performance.

Polymers in self-healing concrete have been also studied in the form of ***shape memory polymers*** (SMPs) which are shrinkable polyethylene terephthalate (PET) filaments. These filaments are shaped in the form of tendons and embedded within a cementitious matrix. When a crack forms, SMPs can be thermally activated which

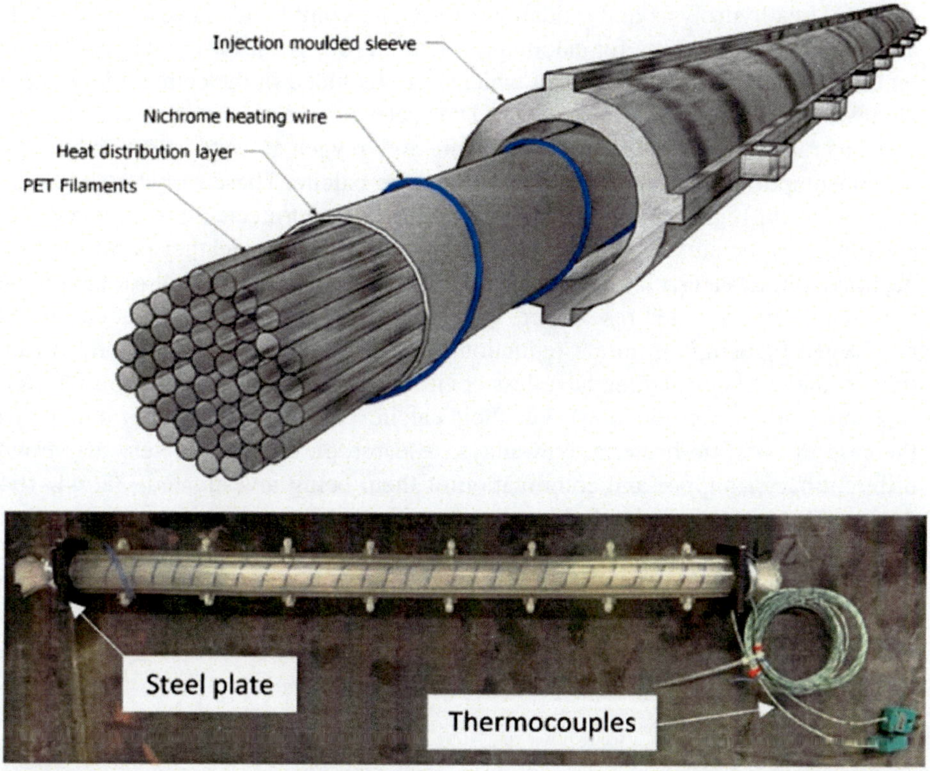

Fig. 6 On the top it is the schematic representation of the updated SMP tendon system and below how it looks prior to installation. Image reproduced from [99]

will make them shrink. The shrinkage action of the SMP tendon exerts a contraction force on the composite forcing the crack to close. Early attempts and proof of concept involved work on different configurations and manual heating of the system using ovens [67, 98]. Improvement of the system led to the development of a protective sleeve for the SMP tendons, the incorporation of steel plates for appropriate anchorage within the cementitious matrix as well as the use of an in situ heating element that can be activated on demand upon the formation of cracks (Fig. 6) [29, 99]. This updated system showed crack closure performance of up to 85% in unreinforced concrete beams and between 26 and 39% for reinforced concrete beams [99].

2.3.4 Bacterial Induced Healing

The use of ***bacteria*** as a potential mechanism for healing in concrete is one of the earliest concepts explored within this area of interest. The idea is based on the fact that there are bacterial species which through their metabolic cycle they produce calcite. Urea hydrolysis and oxidation of organic compounds have been identified as two main mechanisms for calcite production [100, 101]. Early attempts focused on ureolitic and oxidising bacteria which were included in the cementitious matrix alongside with suitable nutrients [102]. The bacteria remain dormant in concrete, and when a crack formed, the influx of moisture and oxygen activates them and through the consumption of nutrients they start producing calcite. These early tests identified three points of high interest: (i) bacteria when used unprotected in concrete did not perform as expected; (ii) when bacterial protected with silica gel they performed well healing the formed cracks and (iii) the existence of oxygen was vital for the process to initiate. The latter led researchers to investigate bacterial strains that do not rely on oxygen focusing on nitrate reduction bacteria [103]. Denitrifying strains function at low levels of oxygen have less environmental impact than ureolitic bacteria (e.g. no ammonia production), but their calcite production is not as much as in the case of ureolitic bacteria. Nowadays, research on this area is very active with different bacteria types and combination of them being investigated [32, 44, 104]. As highlighted from early studies, the protection of bacteria is very important for their functionality. Towards this direction, researchers have been exploring various pathways including microcapsules [34], impregnation of porous media [105] and hydrogels [106].

2.3.5 Nano-additives

The area least explored to date is the area of ***nano-additives***, whether we are referring to nano-minerals (e.g. nano-silica) or other compounds such as carbon nano-tubes and nano-graphite or even graphene. Although nano-minerals (e.g. nano-clays, nano-silica) have been studied extensively on other self-healing applications (e.g. ceramics, polymers), their use in self-healing cementitious composites is extremely limited. Only recently some researchers have drawn their focus on nano-silica in concrete [107, 108] whereas a study incorporated it to promote self-healing actions in asphalt [109]. Silicon-rich nano-minerals can participate in extensive hydration reactions and form hydration products that can heal formed cracks. Due to their size, they can potentially do that even at very small crack sizes while they can also remain dormant in the cementitious matrix for extensive periods of time. Other nano-additions are known to act very well as nucleation sites where hydration products and formed crystals can bind well and form strong lattices [110–112]. These are properties that certainly need attention in the near future.

3 Assessment of Healing

Although self-healing concrete has drawn considerable attention for more than a decade now, still there is no standardised procedures for assessing healing. The major drawback of this is that it hinders the commercialisation of self-healing concrete.

With no widely acceptable standard methods of assessment, it will be extremely difficult to convince engineers and clients to adopt this technology. On the other hand, this lack of standardisation at this research and development growth stage has provided enough flexibility to researchers for exploring a wide variety of methods and techniques. Nonetheless, there are recent studies that focus on round-robin tests and stress the necessity of some sort of standardisation [113, 114]. Standardisation is not only needed for commercial purposes but also for proper comparison between observations from different research groups. In addition, it needs to be highlighted that the majority of techniques used for assessing self-healing, especially the bulk techniques (e.g. sorptivity; chloride diffusion), were developed originally for pristine concrete samples. Having extensive cracking affects the way the test is performed, and it may well be that in the near future new modified techniques have to be developed. The scope of this section is not to present in detail all the documented assessment techniques for self-healing concrete but rather to provide a general layout/framework of assessment procedures.

3.1 Assessment of Mechanical Properties

Almost every single study on self-healing concrete incorporates some sort of assessment of mechanical properties, with the most common ones being compressive and flexural strengths (either 3- or 4-point bend). For compressive strength tests, specimens are being examined to assess the effect of inclusions (e.g. microcapsules) on strength development. Flexural strength is typically used to induce cracking and then following the healing period to identify strength recovery. A critical element in this discussion is the uniformity of crack sizes between samples. This is important not only for strength recovery assessment but also for any other type of assessment where the crack size has a significant impact (e.g. sorption; crack mouth healing). To ensure that cracks are of the same size, and therefore comparable, flexural tests must be performed in deflection control mode using either linear variable differential transformers (LVDTs) or clip-gauges. The use of such sensors will allow the user to determine desirable crack size for each test.

There are two complications in this discussion. The one is that when working with concrete or coarse mortars, crack sizes cannot be controlled very well as their formation depends on the particle size and location of the coarsest aggregate. On the other hand, when working with fine mortars or cement pastes one must consider the quasi-brittleness of these matrices. Forming a crack under flexure in a quasi-brittle matrix may result in the uncontrollable growth of the crack and hence the complete fracture of the sample. For this reason, many researchers recommend the use of some minimal reinforcement in fine mortar/paste samples to prevent unwanted damage [92]. In addition, the creation of an artificial notch prior to testing is also strongly advisable as this will ensure the specific location of the crack formation. Micromechanical testing in the form of dynamic mechanical analysis and micro-indentation has also been used by some studies to study the properties of encapsulation materials and the healed cementitious matrix [36, 76, 90, 115].

3.2 Assessment of Durability Recovery

In laboratory conditions, especially when dealing with mineral healing processes, it is difficult to achieve significant recovery in mechanical properties. The reason is that typically the samples used rarely are larger than 200 mm long with a cross of maximum 50 × 50 mm. A single crack of 200 microns (which is a typical size of crack found in many studies) represents a significant level of damage in such specimen in term of its structural integrity. Nonetheless, when durability recovery comes into discussion things are considerably improved. Hence, durability recovery is a very favourable area of assessment for in-laboratory assessments. Water permeability and water capillary sorption have been the most favourable testing procedures. For water permeability, the needed pressure to promote the permeation mechanism is ensured through some water-pressure head.

As mentioned before, the lack of standardisation has allowed researchers to improvise and develop different configurations depending on their proof of concept and the type and size of their specimens. Capillary sorption is more straightforward, and the test is typically done on a small prism with a crack in the midspan. The bottom side of the sample is covered leaving only the cracked/healed area exposed to water sorption. Sorptivity tests have been proved very robust in defining levels of healing in cement-based composites. Other durability recovery techniques reported so far are rapid chloride permeability [55, 116, 117], chloride ion diffusion [118] and gas permeability [35]. It is evident that work has been mostly one sided on durability recovery and more work is needed to understand how healing processes evolve under other conditions (e.g. carbonation; freeze–thaw) and even in a set of combinatory scenarios.

3.3 Visual Assessment (Meso-/Micro- and Indirect)

Visual assessment of cracks has been a very efficient way to have a first indication of the level of healing. Visual assessment can be at meso-level using stereomicroscopes and at microlevel using optical and/or scanning electron microscopes (OM and SEM). The former is mainly used to assess the healing development at the crack mouth verifying in detail the macroscopic observation of a healed crack (Fig. 7). OM/SEMs are used for assessing the nature and characteristics of the healing products forming within the crack plane. In some cases, elemental analysis is combined with SEM imaging to verify in the exact consistency of the healing products (Fig. 8).

Visual assessment provides a clear indication of healing on the mouth of the crack; it helps to gain good understanding of the nature of the healing products within the

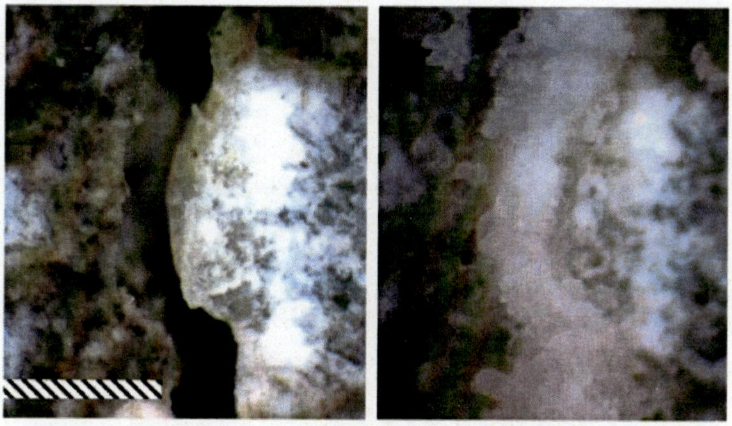

Fig. 7 Representative image of crack mouth healing: **a** the crack mouth on the day of cracking and **b** the healed crack after 28 days [35]. Image reproduced with permission from Elsevier

crack plane. However, it cannot provide an indication of the extent of healing in the bulk volume of the crack. For this reason, indirect visualisation methods such as the ultrasonic pulse velocity measurements (UPV) are used. Using a suitable UPV platform with adequate algorithm for crack depth calculation, one can estimate how much of the crack volume has been filled after the healing period (Fig. 9). Another similar approach to UPV is the evaluation of crack healing in the whole depth of crack using acoustic emission (AE) approach. Although some researchers have utilised AE, the technique has not been used widely, probably due to the specialised equipment required [73, 119].

A more advanced way to have an idea of the healing within the crack plane is the use of X-ray computer tomography (XCT). This technique allows the development of a 3D image of the core of the subject under inspection. Some researchers have already use generating important information about the healing levels, distribution of microcapsules, etc. (Fig. 10) [38, 120, 121].

3.4 *Microstructural Assessment*

Microstructural assessment involves a wide spectrum of techniques used to identify the effect of additions in self-healing matrices and/or to gain better understanding of the nature and type of the formed healing products. SEM and elemental analysis can fall within this category but were discussed earlier. Other methods include isothermal calorimetry, thermogravimetric analysis (TGA), X-ray diffraction and Fourier transform infrared spectroscopy (FTIR). Isothermal calorimetry can be used to describe

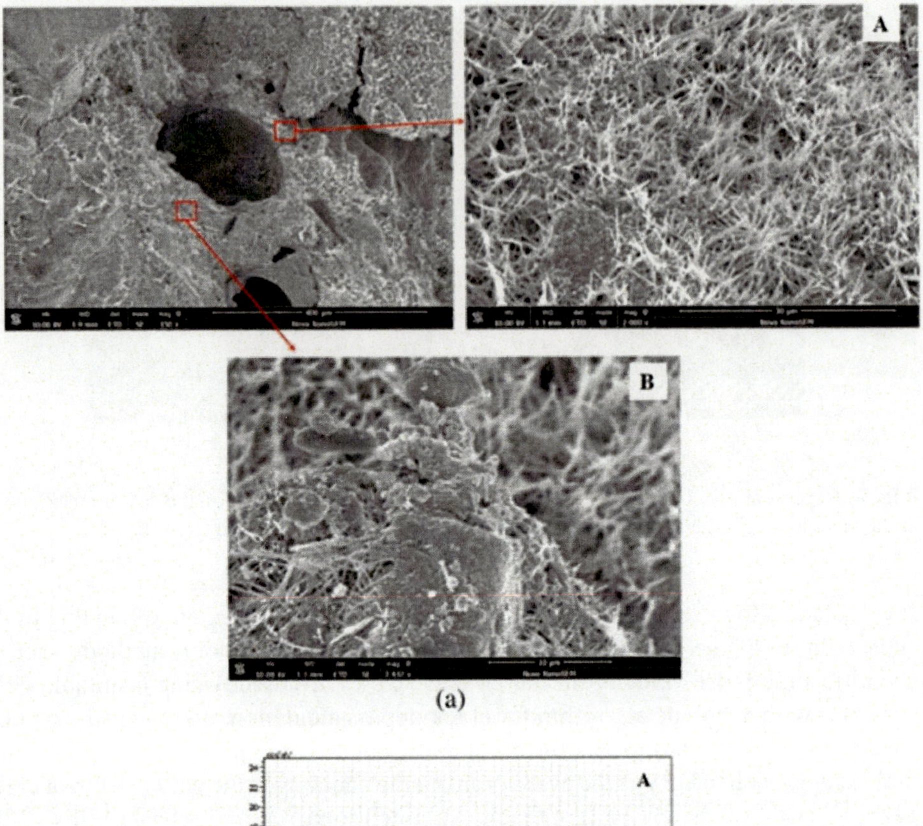

Fig. 8 SEM image depicting the area around a fractured microcapsule and associated elemental analysis of the formed products around this area verifying the existence of Ca- and Si-rich products, which signals the proliferation of healing compounds. Image reproduced from [92]

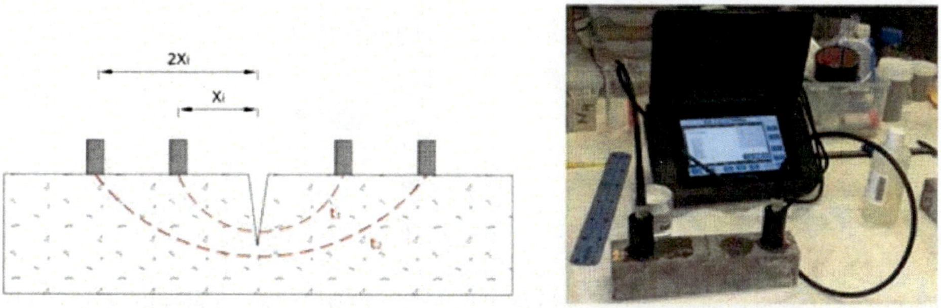

Fig. 9 A typical laboratory UPV set-up for measuring crack depths. Image reproduced from [92]

Fig. 10 X-ray microcomputer tomography image of a cement-based composite containing microcapsules: **a** 3D reconstruction of the matrix showing the crack and microcapsules and **b** 3D reconstruction of the exact location of the microcapsules in the matrix [122]. Image reproduced with permission from Elsevier

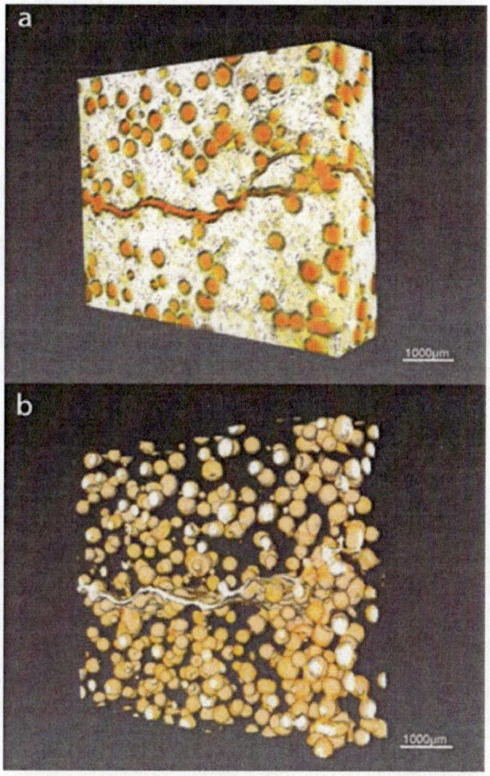

the hydration kinetics of matrices that contain self-healing promoting mineral additions or to explore whether other polymeric additions (e.g. microcapsules) affect the hydration kinetics of the cementitious matrix [89].

TGA, XRD and FTIR are mainly used to characterise healing products extracted from the crack planes. Characterisation may vary from simple identification of phases to calculating the degree of hydration or verifying the existence of carbonate phases. In these techniques, it is very important that the samples are collected very carefully from the crack planes and prepared accordingly for each testing [123]. Collection of

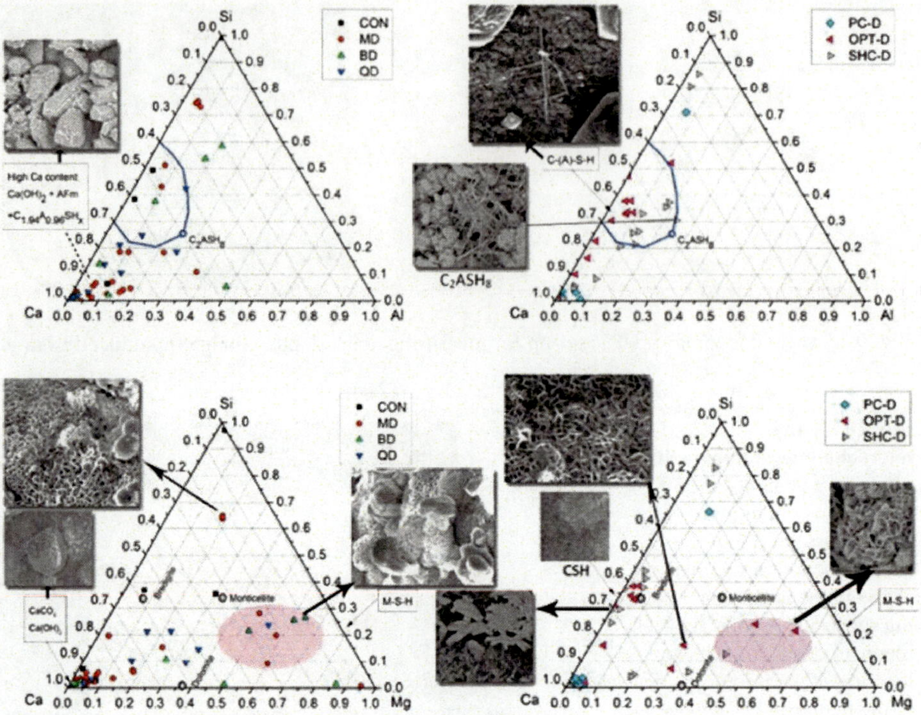

Fig. 11 Ternary diagrams of healing products; plots created using SEM elemental analysis which was verified through FTIR and XRD. Image reproduced from [74]

healing materials is typically done by using a filing tool take care not to contaminate the sample with the rest of the cementitious matrix. The information extracted from such analysis can be combined with data obtained from other techniques (e.g. SEM) and yield to the development of ternary plots which can lead to a more comprehensive understanding of the formed healing materials (Fi. 11).

3.5 *In Situ* Assessment of Healing on Large Scale

Assessing healing in the large scale has several challenges compared to "sterile" environment of a laboratory. First, the exposure conditions are not expected to be constant. This includes both weather-related conditions as well as conditions in the surrounding environment (e.g. vibrations; existence of large pressure or thermal differentials). Secondly, as with standard laboratory testing techniques, some in situ assessment techniques and equipment are not designed to function on cracked and then healed elements. These two parameters make the assessment of healing on the large scale a tricky task with no easy solution.

Besides the complexities of delivering self-healing systems on the large scale, there is gradually growing interest in such applications. Several attempts are reported in the UK, Continental Europe and China [82, 124–126]. The most common approach in field assessment of healing is the macroscopic visual inspection of cracks. Ultrasonic pulse velocity has been used in two cases as well to identify levels of healing [38, 82, 126]. In addition to UPV, researchers in the UK employed other techniques to assess healing of large concrete elements [38, 82]. Digital image correlation was used to identify strength recovery. Permeability recovery was assessed with an in situ air permeameter while healing material was carefully extracted from cracks to perform microstructural characterisation.

References

1. Beng PP (2013) The carbon footprint of reinforced concrete. Adv Cem Res 25:362–368. http://dx.doi.org/https://doi.org/10.1680/adcr.13.00013
2. Building Research Establishment (2003) Residual life models for concrete repair—assessment of the concrete repair process
3. Office for National Statistics (2016) Construction output in great Britain: May 2016. Off Natl Stat 5–9
4. Ferrara L, Krelani V (2013) A fracture testing based approach to assess the self healing capacity of cementitious composites. In: Van Mier JG, Ruiz G, Andrade C et al (eds) VIII international conference on fracture mechanics of concrete and concrete structures. Toledo
5. Abrams D (1913) Tests of bond between concrete and steel
6. Lauer K, Slate F (1956) Autogenous healing of cement paste. ACI J Proc 52:1083–1098. https://doi.org/10.14359/11661
7. Wagner EF (1974) Autogenous healing of cracks in cement-mortar linings for gray-iron an ductile-iron water pipe. J Am Water Work Assoc 66:358–360. https://doi.org/10.1002/j.1551-8833.1974.tb02046.x
8. Ivanov FM, Polyakov BI (1974) Self-healing and durability of hydraulic concrete. Hydrotechnical Constr 8:844–849. https://doi.org/10.1007/BF02380442
9. Gray RJ (1984) Autogenous healing of fibre/matrix interfacial bond in fibre-reinforced mortar. Cem Concr Res 14:315–317. https://doi.org/10.1016/0008-8846(84)90047-4
10. Hannant DJ, Keer JG (1983) Autogenous healing of thin cement based sheets. Cem Concr Res 13:357–365. https://doi.org/10.1016/0008-8846(83)90035-2
11. Zamorowski W (1985) The phenomenon of self-regeneration of concrete. Int J Cem Compos Light Concr 7:199–201. https://doi.org/10.1016/0262-5075(85)90008-9
12. Morlan L (1988) Rehabilitation of lake harriet dam. In: Waterpower '87, proceedings of the International Conference on Hydropower, Portland, OR, USA, Code 11052. ASCE, pp 1344–1352
13. Edvardsen C (1999) Water permeability and autogeneous healing of cracks in concrete. ACI Mater J 96:448–454
14. Reinhardt HW, Jooss M (2003) Permeability and self-healing of cracked concrete as a function of temperature and crack width. Cem Concr Res 33:981–985. https://doi.org/10.1016/S0008-8846(02)01099-2
15. Mihashi H, Kaneko Y, Nishiwaki T, Otsuka K (2000) Fundamental study on development of intelligent concrete characterized by self-healing capability for strength. Trans Japan Concr Inst 22:441–450. https://doi.org/10.3151/crt1990.11.2_21
16. Jonkers HM, Schlangen E (2007) Self-healing of cracked concrete: A bacterial approach. In: Proceedings of the 6th international conference on fracture mechanics of concrete and concrete structures. pp 1821–1826

17. Wiktor V, Jonkers HM (2011) Quantification of crack-healing in novel bacteria-based self-healing concrete. Cem Concr Compos 33:763–770. https://doi.org/10.1016/j.cemconcomp.2011.03.012
18. Homma D, Mihashi H, Nishiwaki T (2009) Self-healing capability of fibre reinforced cementitious composites. J Adv Concr Technol 7:217–228. https://doi.org/10.3151/jact.7.217
19. Koda M, Mihashi H, Nishiwaki T, Kikuta T (2011) Experimental study on self-healing capability of frcc using synthetic fibers. J Struct Constr Eng 76:1547–1552. https://doi.org/10.3130/aijs.76.1547
20. Zhou Z, Li Z, Xu D, Yu J (2011) Influence of slag and fly ash on the self-healing ability of concrete. In: Advanced materials research. pp 1020–1023
21. Watanabe T, Fujiwara Y, Hashimoto C, Ishimaru K (2011) Evaluation of self healing effect in fly-ash concrete by ultrasonic test method. In: International J Mod Phys B, pp 4307–4310
22. De Belie N, Van Tittelboom K, Snoeck D, Wang J (2012) Smart additives for self-sealing and self-healing concrete. In: Materials research society symposium proceedings. Materials Research Society, pp 129–140
23. Yang Z, Hollar J, He X, Shi X (2011) A self-healing cementitious composite using oil core/silica gel shell microcapsules. Cem Concr Compos 33:506–512. https://doi.org/10.1016/j.cemconcomp.2011.01.010
24. Gilford J III, Hassan MM, Rupnow T et al (2014) Dicyclopentadiene and Sodium Silicate Microencapsulation for Self-Healing of Concrete. J Mater Civ Eng 26:886–896. https://doi.org/10.1061/(ASCE)MT.1943-5533.0000892
25. Jefferson A, Joseph C, Lark R et al (2010) A new system for crack closure of cementitious materials using shrinkable polymers. Cem Concr Res 40:795–801. https://doi.org/10.1016/j.cemconres.2010.01.004
26. Ferrara L, Krelani V, Carsana M (2014) A "' fracture testing "' based approach to assess crack healing of concrete with and without crystalline admixtures. Constr Build Mater 68:535–551. https://doi.org/10.1016/j.conbuildmat.2014.07.008
27. Roig-Flores M, Moscato S, Serna P, Ferrara L (2015) Self-healing capability of concrete with crystalline admixtures in different environments. Constr Build Mater 86:1–11. https://doi.org/10.1016/j.conbuildmat.2015.03.091
28. Qureshi T, Kanellopoulos A, Al-Tabbaa A (2016) Encapsulation of expansive powder minerals within a concentric glass capsule system for self-healing concrete. Constr Build Mater 121:629–643. https://doi.org/10.1016/j.conbuildmat.2016.06.030
29. Teall OR, Pilegis M, Sweeney J et al (2017) Development of high shrinkage Polyethylene Terephthalate (PET) shape memory polymer tendons for concrete crack closure. Smart Mater Struct 26:045006. https://doi.org/10.1088/1361-665X/aa5d66
30. Li G, Zhang P (2013) A self-healing particulate composite reinforced with strain hardened short shape memory polymer fibers. Polymer (Guildf) 54:5075–5086. https://doi.org/10.1016/j.polymer.2013.07.010
31. Zhang JL, Wang CG, Wang QL et al (2016) A binary concrete crack self-healing system containing oxygen-releasing tablet and bacteria and its Ca2+-precipitation performance. Appl Microbiol Biotechnol. https://doi.org/10.1007/s00253-016-7741-z
32. Palin D, Wiktor V, Jonkers HM (2016) A bacteria-based bead for possible self-healing marine concrete applications. Smart Mater Struct 25:1–6. https://doi.org/10.1088/0964-1726/25/8/084008
33. Tziviloglou E, Wiktor V, Jonkers HM, Schlangen E (2016) Bacteria-based self-healing concrete to increase liquid tightness of cracks. Constr Build Mater 122:118–125. https://doi.org/10.1016/j.conbuildmat.2016.06.080
34. Wang JY, Soens H, Verstraete W, De Belie N (2014) Self-healing concrete by use of microencapsulated bacterial spores. Cem Concr Res 56:139–152. https://doi.org/10.1016/j.cemconres.2013.11.009
35. Kanellopoulos A, Qureshi TS, Al-Tabbaa A (2015) Glass encapsulated minerals for self-healing in cement based composites. Constr Build Mater 98:780–791. https://doi.org/10.1016/j.conbuildmat.2015.08.127

36. Kanellopoulos A, Giannaros P, Al-Tabbaa A (2016) Polymeric microcapsules with switchable mechanical properties for self-healing concrete—synthesis and proof of concept. Smart Mater Struct 26:045025. http://dx.doi.org/https://doi.org/10.1088/1361-665X/aa516c
37. RM4L M& (2021) M4L & RM4L. https://rm4l.com/. Accessed 9 Apr 2021
38. Al-Tabbaa A, Litina C, Giannaros P et al (2019) First UK field application and performance of microcapsule-based self-healing concrete. Constr Build Mater 208. https://doi.org/10.1016/j.conbuildmat.2019.02.178
39. Basilik (2021) Green basilisk makes self-healing concrete using Bacteria. https://www.labiotech.eu/startup-scout/green-basilisk-self-healing-concrete/. Accessed 9 Apr 2021
40. Wu M, Johannesson B, Geiker M (2012) A review: Self-healing in cementitious materials and engineered cementitious composite as a self-healing material. Constr Build Mater 28:571–583. https://doi.org/10.1016/j.conbuildmat.2011.08.086
41. Mihashi H, Nishiwaki T (2012) Development of engineered self-healing and self-repairing concrete-state-of-the-art report. J Adv Concr Technol 10:170–184. https://doi.org/10.3151/jact.10.170
42. Van Tittelboom K, De Belie N (2013) Self-healing in cementitious materials—a review. Materials (Basel) 6:2182–2217. https://doi.org/10.3390/ma6062182
43. de Rooij MR, Tittelboom K Van, De Belie N, Schlangen E (2013) Self-healing phenomena in cement-based materials: state-of-the-art report of RILEM technical committee 221-SHC: self-healing phenomena in cement-based materials. Springer
44. De BN, Gruyaert E, Al-tabbaa A et al (2018) A review of self-healing concrete for damage management of structures. Adv Mater Interfaces 1800074:1–28. https://doi.org/10.1002/admi.201800074
45. Kan L, Shi H (2012) Investigation of self-healing behavior of Engineered Cementitious Composites (ECC) materials. Constr Build Mater 29:348–356. https://doi.org/10.1016/j.conbuildmat.2011.10.051
46. Snoeck D, Tittelboom KV, Steuperaert S et al (2012) Self-healing cementitious materials by the combination of microfibres and superabsorbent polymers. J Intell Mater Syst Struct. https://doi.org/10.1177/1045389X12438623
47. Lothenbach B, Scrivener K, Hooton RD (2011) Supplementary cementitious materials. Cem Concr Res 41:1244–1256
48. Neville AM (2011) Properties of concrete, 5th edn. Prentice Hall
49. Hewlett PC, Liska M (2019) Lea's chemistry of cement and concrete. Elsevier
50. Termkhajornkit P, Nawa T, Yamashiro Y, Saito T (2009) Self-healing ability of fly ash-cement systems. Cem Concr Compos 31:195–203. https://doi.org/10.1016/j.cemconcomp.2008.12.009
51. Hung CC, Su YF (2016) Medium-term self-healing evaluation of engineered cementitious composites with varying amounts of fly ash and exposure durations. Constr Build Mater 118:194–203. https://doi.org/10.1016/j.conbuildmat.2016.05.021
52. Qian S, Zhou J, de Rooij MR et al (2009) Self-healing behavior of strain hardening cementitious composites incorporating local waste materials. Cem Concr Compos 31:613–621. https://doi.org/10.1016/j.cemconcomp.2009.03.003
53. Van Tittelboom K, Gruyaert E, Rahier H, De Belie N (2012) Influence of mix composition on the extent of autogenous crack healing by continued hydration or calcium carbonate formation. Constr Build Mater 37:349–359. https://doi.org/10.1016/j.conbuildmat.2012.07.026
54. Huang H, Ye G, Damidot D (2013) Characterization and quantification of self-healing behaviors of microcracks due to further hydration in cement paste. Cem Concr Res 52:71–81. https://doi.org/10.1016/j.cemconres.2013.05.003
55. Sahmaran M, Yildirim G, Erdem TK (2013) Self-healing capability of cementitious composites incorporating different supplementary cementitious materials. Cem Concr Compos 35:89–101. https://doi.org/10.1016/j.cemconcomp.2012.08.013
56. Ahn T-H, Kishi T (2010) Crack self-healing behavior of cementitious composites incorporating various mineral admixtures. J Adv Concr Technol 8:171–186. https://doi.org/10.3151/jact.8.171

57. Jiang Z, Li W, Yuan Z (2015) Influence of mineral additives and environmental conditions on the self-healing capabilities of cementitious materials. Cem Concr Compos 57:116–127. https://doi.org/10.1016/j.cemconcomp.2014.11.014
58. Mo L, Deng M, Tang M (2010) Effects of calcination condition on expansion property of MgO-type expansive agent used in cement-based materials. Cem Concr Res 40:437–446. https://doi.org/10.1016/j.cemconres.2009.09.025
59. Qureshi T, Kanellopoulos A, Al-Tabbaa A (2018) Autogenous self-healing of cement with expansive minerals-I: impact in early age crack healing. Constr Build Mater 192:768–784. https://doi.org/10.1016/j.conbuildmat.2018.10.143
60. Qureshi T, Kanellopoulos A, Al-Tabbaa A (2019) Autogenous self-healing of cement with expansive minerals-II: impact of age and the role of optimised expansive minerals in healing performance. Constr Build Mater 194. https://doi.org/10.1016/j.conbuildmat.2018.11.027
61. Roig-Flores M, Pirritano F, Serna P, Ferrara L (2016) Effect of crystalline admixtures on the self-healing capability of early-age concrete studied by means of permeability and crack closing tests. Constr Build Mater 114:447–457. https://doi.org/10.1016/j.conbuildmat.2016.03.196
62. Ferrara L, Krelani V, Moretti F (2016) On the use of crystalline admixtures in cement based construction materials: from porosity reducers to promoters of self healing. Smart Mater Struct 25:084002. https://doi.org/10.1088/0964-1726/25/8/084002
63. S, Wang R, He P et al (2021) Synergistic effect of ion chelating agent and inorganic compound on pore structure, mechanical and self-healing performance of cement-based materials. Smart Mater Struct 30:015011. https://doi.org/10.1088/1361-665X/abc66a
64. Dry C (1994) Smart multiphase composite materials which repair themselves by a release of liquids which become solids. In: Proceedings of symposium on smart structures and materials. Orlando, USA, pp 62–70
65. White SR, Sottos NR, Geubelle PH et al (2001) Autonomic healing of polymer composites. Nature 409:794–797. https://doi.org/10.1038/35057232
66. Toohey KS, Sottos NR, Lewis JA et al (2007) Self-healing materials with microvascular networks. Nat Mater 6:581–5. https://doi.org/10.1038/nmat1934
67. Joseph C, Lark R, Isaacs B et al (2010) Experimental investigation of adhesive-based self-healing of cementitious materials. Mag Concr Res 62:831–843. https://doi.org/10.1680/macr.2010.62.11.831
68. De Belie N, Van Tittelboom K (2010) Self-healing concrete: suitability of different healing agents. Int J 3 R's 1:12–21
69. Dry C, McMillan W (1996) Three-part methylmethacrylate adhesive system as an internal delivery system for smart responsive concrete. Smart Mater Struct 5:297–300. https://doi.org/10.1088/0964-1726/5/3/007
70. Li VC, Lim YM, Chan Y-W (1998) Feasibility study of a passive smart self-healing cementitious composite. Compos Part B Eng 29:819–827. https://doi.org/10.1016/S1359-8368(98)00034-1
71. Joseph C, Jefferson AD, Cantoni MB (2007) Issues relating to the autonomic healing of cementitious materials. In: Proceedings of the first international conference on self healing materials (CDROM). Delft, The Netherlands, pp 1–8
72. Feiteira J, Gruyaert E, De Belie N (2016) Self-healing of moving cracks in concrete by means of encapsulated polymer precursors. Constr Build Mater 102:671–678. https://doi.org/10.1016/j.conbuildmat.2015.10.192
73. Van Tittelboom K, Wang J, Araújo M et al (2016) Comparison of different approaches for self-healing concrete in a large-scale lab test. Constr Build Mater 107:125–137. https://doi.org/10.1016/j.conbuildmat.2015.12.186
74. Qureshi T, Kanellopoulos A, Al-Tabbaa A (2016) Encapsulation of expansive powder minerals within a concentric glass capsule system for self-healing concrete. Constr Build Mater Press
75. Formia A, Irico S, Bertola F et al (2016) Experimental analysis of self-healing cement-based materials incorporating extruded cementitious hollow tubes. J Intell Mater Syst Struct 27:2633–2652. https://doi.org/10.1177/1045389X16635847

76. Hilloulin B, Van Tittelboom K, Gruyaert E et al (2015) Design of polymeric capsules for self-healing concrete. Cem Concr Compos 55:298–307. https://doi.org/10.1016/j.cemconcomp.2014.09.022
77. Šavija B, Feiteira J, Araújo M et al (2017) Simulation-aided design of tubular polymeric capsules for self-healing concrete. Materials (Basel) 10. https://doi.org/10.3390/ma10010010
78. Sisomphon K, Copuroglu O, Fraaij A (2011) Application of encapsulated lightweight aggregate impregnated with sodium monofluorophosphate as a self-healing agent in blast furnace slag mortar. Heron 56:13–32
79. Alghamri R, Kanellopoulos A, Al-Tabbaa A (2016) Impregnation and encapsulation of lightweight aggregates for self-healing concrete. Constr Build Mater 124:910–921. https://doi.org/10.1016/j.conbuildmat.2016.07.143
80. Wang X, Li W, Jiang Z (2020) Preparation and characterization of self-healing mortar based on "build-in" carbonation. Materials (Basel) 13. https://doi.org/10.3390/ma13030644
81. Minnebo P, Thierens G, De Valck G et al (2017) A novel design of autonomously healed concrete: towards a vascular healing network. Materials (Basel) 10:49. https://doi.org/10.3390/ma10010049
82. Davies R, Teall O, Pilegis M et al (2018) Large scale application of self-healing concrete: design, construction, and testing. Front Mater 5:1–12. https://doi.org/10.3389/fmats.2018.00051
83. Davies R, Jefferson A, Lark R, Gardner D (2015) A novel 2D vascular network in cementitious materials. In: Concrete—innovation and design, fib Symposium
84. Selvarajoo T, Davies RE, Gardner DR, et al (2020) Characterisation of a vascular self-healing cementitious material system: flow and curing properties. Constr Build Mater 245:118332. https://doi.org/10.1016/j.conbuildmat.2020.118332
85. Li Z, Souza LR de, Litina C et al (2020) A novel biomimetic design of a 3D vascular structure for self-healing in cementitious materials using Murray's law. Mater Des 190:108572. https://doi.org/10.1016/j.matdes.2020.108572
86. Pelletier M, Bose A (2011) Self-mending composites incorporating encapsulated mending agents. 1
87. Dong B, Wang Y, Fang G et al (2015) Smart releasing behavior of a chemical self-healing microcapsule in the stimulated concrete pore solution. Cem Concr Compos 56:46–50. https://doi.org/10.1016/j.cemconcomp.2014.10.006
88. Perez G, Erkizia E, Gaitero JJ et al (2015) Synthesis and characterization of epoxy encapsulating silica microcapsules and amine functionalized silica nanoparticles for development of an innovative self-healing concrete. Mater Chem Phys 165:39–48. https://doi.org/10.1016/j.matchemphys.2015.08.047
89. Giannaros P, Kanellopoulos A, Al-Tabbaa A (2016) Sealing of cracks in cement using microencapsulated sodium silicate. Smart Mater Struct 25:84005. https://doi.org/10.1088/0964-1726/25/8/084005
90. Souza L, Al-Tabbaa A (2018) Microfluidic fabrication of microcapsules tailored for self-healing in cementitious materials. Constr Build Mater 184:713–722. https://doi.org/10.1016/j.conbuildmat.2018.07.005
91. de Souza LR, Al-Tabbaa A, Rossi D (2019) Taking a microfluidic approach to the production of self-healing construction materials. Met Powder Rep 74:121–125. https://doi.org/10.1016/j.mprp.2019.01.001
92. Kanellopoulos A, Giannaros P, Al-Tabbaa A (2016) The effect of varying volume fraction of microcapsules on fresh, mechanical and self-healing properties of mortars. Constr Build Mater 122:577–593. https://doi.org/10.1016/j.conbuildmat.2016.06.119
93. Mechtcherine V, Reinhardt HW (2012) STAR 225-SAP application of superabsorbent polymers (SAP) in concrete construction. Springer
94. Lee HXD, Wong HS, Buenfeld NR (2015) Self-sealing of cracks in concrete using superabsorbent polymers. Cem Concr Res 79:194–208. https://doi.org/10.1016/j.cemconres.2015.09.008

95. Park B, Choi YC (2018) Self-healing capability of cementitious materials with crystalline admixtures and super absorbent polymers (SAPs). Constr Build Mater 189:1054–1066. https://doi.org/10.1016/j.conbuildmat.2018.09.061
96. Chindasiriphan P, Yokota H, Pimpakan P (2020) Effect of fly ash and superabsorbent polymer on concrete self-healing ability. Constr Build Mater 233:116975. https://doi.org/10.1016/j.conbuildmat.2019.116975
97. Baloch H, Usman M, Rizwan SA, Hanif A (2019) Properties enhancement of super absorbent polymer (SAP) incorporated self-compacting cement pastes modified by nano silica (NS) addition. Constr Build Mater 203:18–26. https://doi.org/10.1016/j.conbuildmat.2019.01.096
98. Isaacs B, Lark R, Jefferson T et al (2013) Crack healing of cementitious materials using shrinkable polymer tendons. Struct Concr 14:138–147. https://doi.org/10.1002/suco.201200013
99. Teall O, Pilegis M, Davies R et al (2018) A shape memory polymer concrete crack closure system activated by electrical current. Smart Mater Struct 27:075016. https://doi.org/10.1088/1361-665X/aac28a
100. De Muynck W, De Belie N, Verstraete W (2010) Microbial carbonate precipitation in construction materials: a review. Ecol Eng 36:118–136. https://doi.org/10.1016/j.ecoleng.2009.02.006
101. Jonkers HM, Thijssen A, Muyzer G et al (2010) Application of bacteria as self-healing agent for the development of sustainable concrete. Ecol Eng 36:230–235. https://doi.org/10.1016/j.ecoleng.2008.12.036
102. Van Tittelboom K, De Belie N, De Muynck W, Verstraete W (2010) Use of bacteria to repair cracks in concrete. Cem Concr Res 40:157–166. https://doi.org/10.1016/j.cemconres.2009.08.025
103. Erşan YÇ, Hernandez-Sanabria E, Boon N, De Belie N (2016) Enhanced crack closure performance of microbial mortar through nitrate reduction. Cem Concr Compos 70:159–170. https://doi.org/10.1016/j.cemconcomp.2016.04.001
104. Sharma TK, Alazhari M, Heath A et al (2017) Alkaliphilic bacillus species show potential application in concrete crack repair by virtue of rapid spore production and germination then extracellular calcite formation. J Appl Microbiol 122:1233–1244. https://doi.org/10.1111/jam.13421
105. Alazhari M, Sharma T, Heath A et al (2018) Application of expanded perlite encapsulated bacteria and growth media for self-healing concrete. Constr Build Mater 160:610–619. https://doi.org/10.1016/j.conbuildmat.2017.11.086
106. Wang J, Mignon A, Snoeck D et al (2015) Application of modified-alginate encapsulated carbonate producing bacteria in concrete: a promising strategy for crack self-healing. Front Microbiol 6. https://doi.org/10.3389/fmicb.2015.01088
107. Lefever G, Snoeck D, Aggelis DG et al (2020) Evaluation of the self-healing ability of mortar mixtures containing superabsorbent polymers and nanosilica. Materials (Basel) 13:380. https://doi.org/10.3390/ma13020380
108. Akın A (2021) Investigation of different permeability properties of self-healing cementitious composites under colloidal nano silica curing conditions. Struct Concr. https://doi.org/10.1002/suco.202000468
109. Ganjei MA, Aflaki E (2019) Application of nano-silica and styrene-butadiene-styrene to improve asphalt mixture self healing. Int J Pavement Eng 20:89–99. https://doi.org/10.1080/10298436.2016.1260130
110. Han B, Zhang L, Zeng S et al (2017) Nano-core effect in nano-engineered cementitious composites. Compos Part A Appl Sci Manuf 95:100–109. https://doi.org/10.1016/j.compositesa.2017.01.008
111. Sanchez F, Sobolev K (2010) Nanotechnology in concrete—a review. Constr Build Mater 24:2060–2071. https://doi.org/10.1016/j.conbuildmat.2010.03.014
112. Wang B, Jiang R, Wu Z (2016) Investigation of the mechanical properties and microstructure of graphene nanoplatelet-cement composite. Nanomaterials 6:200. https://doi.org/10.3390/nano6110200

113. Litina C, Bumanis G, Anglani G et al (2021) Evaluation of methodologies for assessing self-healing performance of concrete with mineral expansive agents: an interlaboratory study. Materials (Basel) 14:2024. https://doi.org/10.3390/ma14082024
114. Van Mullem T, Anglani G, Dudek M et al (2020) Addressing the need for standardization of test methods for self-healing concrete: an inter-laboratory study on concrete with macrocapsules. Sci Technol Adv Mater 21:661–682. https://doi.org/10.1080/14686996.2020.1814117
115. Xu J, Yao W (2014) Multiscale mechanical quantification of self-healing concrete incorporating non-ureolytic bacteria-based healing agent. Cem Concr Res 64:1–10. https://doi.org/10.1016/j.cemconres.2014.06.003
116. Yıldırım G, Khiavi AH, Yeşilmen S, Şahmaran M (2018) Self-healing performance of aged cementitious composites. Cem Concr Compos 87:172–186. https://doi.org/10.1016/j.cemconcomp.2018.01.004
117. Dong B, Fang G, Ding W et al (2016) Self-healing features in cementitious material with urea—formaldehyde/epoxy microcapsules. Constr Build Mater 106:608–617. https://doi.org/10.1016/j.conbuildmat.2015.12.140
118. Maes M, Van Tittelboom K, De Belie N (2014) The efficiency of self-healing cementitious materials by means of encapsulated polyurethane in chloride containing environments. Constr Build Mater 71:528–537. https://doi.org/10.1016/j.conbuildmat.2014.08.053
119. Granger S, Loukili A, Pijaudier-Cabot G, Chanvillard G (2007) Experimental characterization of the self-healing of cracks in an ultra high performance cementitious material: mechanical tests and acoustic emission analysis. Cem Concr Res 37:519–527. https://doi.org/10.1016/j.cemconres.2006.12.005
120. Sangadji S, Schlangen E (2013) Mimicking bone healing process to self repair concrete structure novel approach using porous network concrete. Procedia Eng 54:315–326. https://doi.org/10.1016/j.proeng.2013.03.029
121. Van Tittelboom K, De Belie N, Van Loo D, Jacobs P (2011) Self-healing efficiency of cementitious materials containing tubular capsules filled with healing agent. Cem Concr Compos 33:497–505. https://doi.org/10.1016/j.cemconcomp.2011.01.004
122. Lv L, Yang Z, Chen G et al (2016) Synthesis and characterization of a new polymeric microcapsule and feasibility investigation in self-healing cementitious materials. Constr Build Mater 105:487–495. https://doi.org/10.1016/j.conbuildmat.2015.12.185
123. Scrivener K, Snellings R, Lothenbach B (2016) A practical guide to microstructural analysis of cementitious materials
124. Van MT, Gruyaert E, Caspeele R, De BN (2020) First large scale application with self-healing concrete in Belgium: analysis of the laboratory control tests. Materials (Basel) 13:997. https://doi.org/10.3390/ma13040997
125. Wang X, Huang Y, Huang Y et al (2019) Laboratory and field study on the performance of microcapsule-based self-healing concrete in tunnel engineering. Constr Build Mater 220:90–101. https://doi.org/10.1016/j.conbuildmat.2019.06.017
126. Qian C, Zheng T, Zhang X, Su Y (2021) Application of microbial self-healing concrete: Case study. Constr Build Mater 290:123226. https://doi.org/10.1016/j.conbuildmat.2021.123226

Self-Healing in Metal-Based Systems

Mariia Arseenko, Julie Gheysen, Florent Hannard, Nicolas Nothomb, and Aude Simar

Abstract Self-healing technologies in metals have a great potential to improve structures reliability and sustainability in particular in the construction sector. However, limited technologies are available compared to other self-healing material groups and they struggle to find industrial applications. The main limitation for self-healing strategies in metals is the low mobility of atoms at room temperature and often the need for an external driving force to promote mass transfer. This chapter provides a review of all currently developed self-healing concepts in metallic systems classified by their scale and healing process category as: macroscopic systems (including liquid-based systems, shape memory alloys and electro-healing) and solid-state healing of nano and microscale damage. A summary of all strategies is provided and a comprehensive analysis of their advantages and challenges is introduced. Finally, further perspectives of self-healing strategies are discussed and critical points interrupting further development of the existing self-healing materials are highlighted.

M. Arseenko · J. Gheysen · F. Hannard · N. Nothomb · A. Simar (✉)
Institute of Mechanics, Materials and Civil Engineering, Université Catholique de Louvain, IMAP, B-1348 Louvain-la-Neuve, Belgium
e-mail: aude.simar@uclouvain.be

M. Arseenko
e-mail: mariia.arseenko@uclouvain.be

J. Gheysen
e-mail: julie.gheysen@uclouvain.be

F. Hannard
e-mail: florent.hannard@uclouvain.be

N. Nothomb
e-mail: nicolas.nothomb@uclouvain.be

© Springer Nature Switzerland AG 2022

A. Kanellopoulos and J. Norambuena-Contreras (eds.), *Self-Healing Construction Materials*, Engineering Materials and Processes,
https://doi.org/10.1007/978-3-030-86880-2_3

1 Introduction

Metallic materials, such as steels and aluminium alloys representing the largest fraction of metal use by volume [1], have enabled technological progress in key sectors such as construction, energy and transportation. With the advent of high-performance structural metals, such as high-strength aluminium alloys, the demand is increasing and the production is predicted to grow by rates of up to 200% until 2050 [1]. However, the production and manufacturing of these metals require a lot of energy and emit pollution. Research around the world aims to develop methods of improving their sustainability. Self-healing metals, and their associated improved alloy longevity, constitute one promising technology among others [1].

Furthermore, self-healing metals are of great interest for (i) applications where structural properties have to be guaranteed, even in presence of small and undetectable damage (e.g. aircraft, space), (ii) applications where a very long life is to be guaranteed, (iii) applications where repair to local damage is costly and disruptive [2], (iv) applications include components that are subjected to infrequent service, require high reliability, or operate in remote conditions (e.g. irradiated environment, in space). The most advanced man-made self-healing materials are polymer-based systems [3, 4], but re-establishing the secondary bonds in between polymeric chains is a lesser challenge than re-establishing metallic bonds. Indeed, chemical reactions in polymers are very efficient in producing a significant release in energy compared to their typical bonding strength. In addition, relatively fast and massive diffusional processes are feasible in polymers even at room temperature [5–7]. The main difficulty with self-healing metallic systems is the low mobility of atoms at room temperature [6, 8–10] and often the need for an external driving force to promote mass transfer [6, 8–12], without sacrificing the strength of the system.

Self-healing metals can be referenced as *autonomous* or *non-autonomous* depending on the need for external human intervention (typically a thermal treatment) [13]. Healing can happen at nano/micro or macroscopic scale. Nano/microscopic self-healing metals are the only autonomous systems. They are based on the activation of metallic atoms diffusion, i.e. activation of the healing agents mobility, in order to fill nano-scale voids and prevent macro-scale damage [13]. Macroscopic scale healing is based on liquid metal flow or electro-healing. A structural macroscopic self-healing metal would be particularly advantageous in applications where part replacement is difficult or impossible [14]. All metal-based self-healing systems currently reported in literature are schematised in Fig. 1.

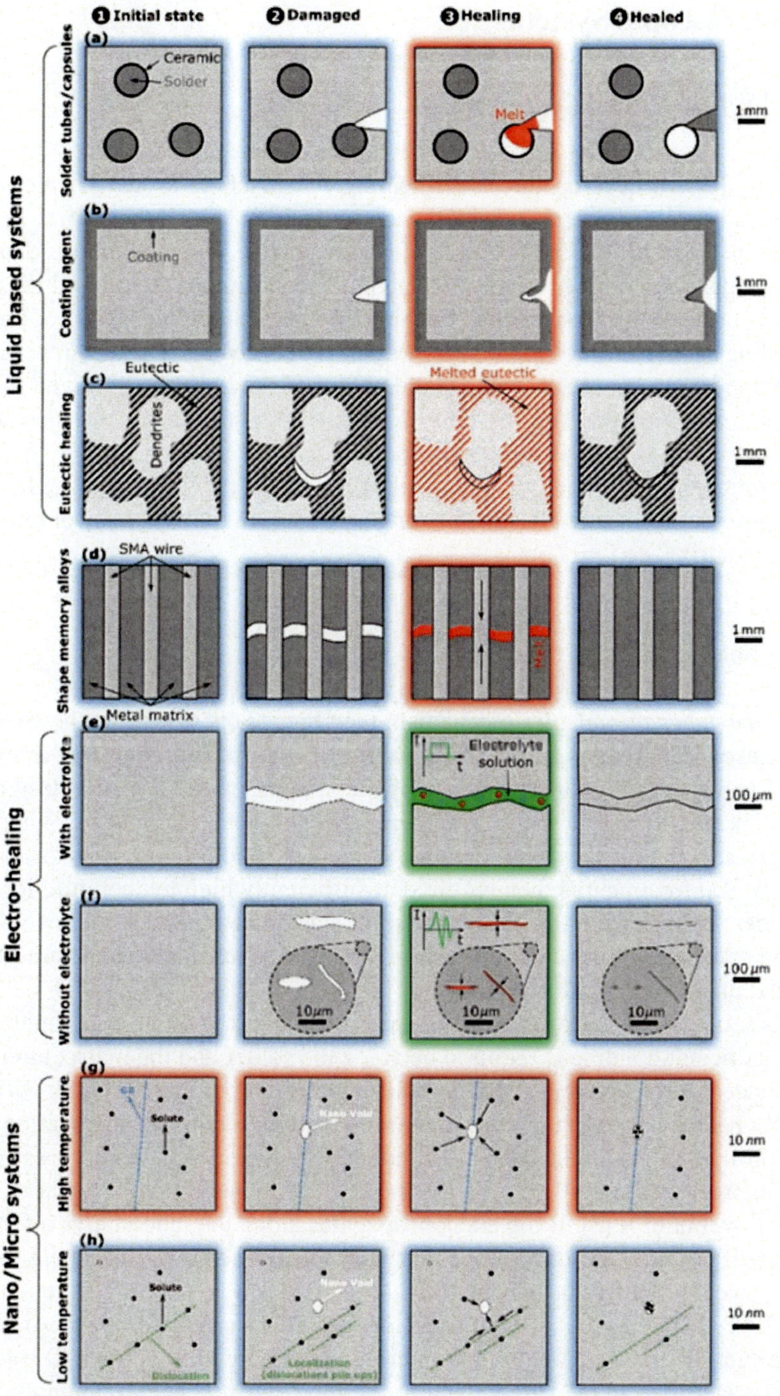

Fig. 1 Overview of the different healing strategies developed for self-healing in metal-based systems. Inspired from [6]

2 Macroscopic Systems

2.1 *Liquid-Based Systems*

This section will develop three self-healing macroscopic systems which may be categorised as *liquid based*. Indeed, the common basis behind these three systems is that they use a liquid healing agent to close the cracks. When the part is damaged, a heat treatment is required to melt the healing agent without melting all the metallic system. The healing agent can then flow into the crack and fill it. After cooling down, the healing agent solidifies in the crack which is thus healed. The difference between these three liquid-based strategies is the nature of the healing agent and the way it will reach the damaged site [6]. This section will first describe the mechanism in which the healing agent is encapsulated (see Fig. 1a). Then, a strategy involving a coating containing the healing agent will be presented (see Fig. 1b). The last system is composed of a eutectic phase as healing agent which melts at a lower temperature than the matrix (see Fig. 1c).

2.1.1 Solder Tubes or Capsules

The *solder tubes or capsules* mechanism (see Fig. 1a) is similar to the well-known system used in polymers [3]. It consists of a low-temperature melting alloy encapsulated inside ceramic tubes or capsules, depending on the selected design. These containers are distributed within a higher melting temperature matrix. When a crack propagates in the material, the ceramic container should break first for healing to apply. A heat treatment is then required at sufficiently high temperature for melting the solder while keeping the matrix in solid state. The melted solder can thus flow into the crack exploiting capillary action and surface tension. As a result, the crack is sealed upon solidification [15, 16].

This self-healing strategy faces two main challenges. First, to completely seal the crack, sufficient bonding is required between the solder and the matrix after healing [16]. Indeed, if the bonding is not complete and is too porous, the crack will only be partially healed and might continue its propagation. Secondly, this strategy totally relies on fracture of the healing agent containers. The crack should break the tube or the microcapsule in order to release the healing agent otherwise the healing agent cannot flow out of it [6]. If the crack propagates along the tube/matrix interface, the solder will not flow out of the tube and will not fill the crack upon heat treatment. Healing would then not be achievable.

In 2008, Martinez Lucci et al. developed a self-healing composite made of an aluminium 206 matrix and alumina microtubes filled with a Sn60Pb40 solder [16]. A hole was drilled in the surface and in the tube. Then, a heat treatment at 300 °C for 5 min was applied on the self-healing composite allowing the solder to flow into this hole and seal it partially.

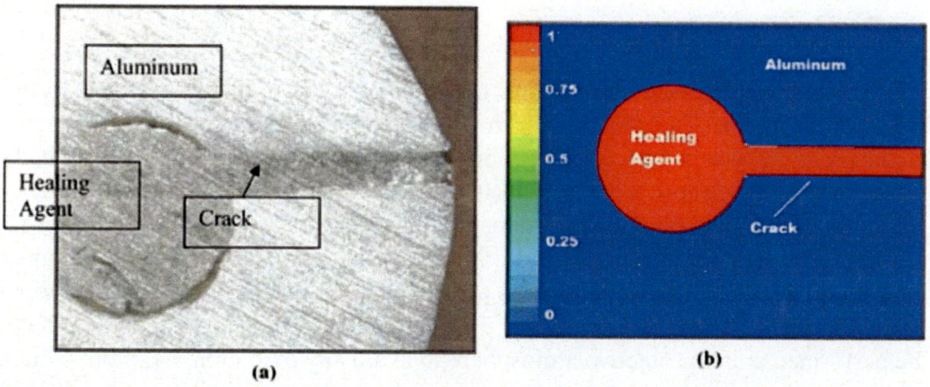

Fig. 2 Healing of crack in a Sn/Al system: **a** experimental result and **b** numerical simulation, the colour code reveals the healing agent fraction. Reproduced from [15]

In 2017, Martinez Lucci et al. investigated the fluid flow and heat transfer occurring in this healing mechanism to investigate its self-healing property [15]. Based on their numerical simulations, they demonstrated the feasibility to design such self-healing composites and that the zinc/aluminium, bismuth/aluminium, indium/tin and tin/tin–bismuth systems present self-healing potential. Figure 2 shows the experimental and numerical results of the healing of a crack which fractures a tube filled with Sn and located inside an Al matrix [15]. The healing agent crack filling is well predicted by the model. However, the interface between the matrix and the healing agent is weak, i.e. with no metallurgical bonding. Indeed, to design a self-healing composite, crucial parameters such as wetting, capillary pressure, viscosity and solidification need to be tailored. Following this purpose, Martinez Lucci et al. defined a healing factor Hf, see Eq. (1). The numerator is composed of the parameters which prevent the flow of the healing agent into the crack, while the denominator is composed of the parameters which promote its flow. It means that when the value of Hf is high, the flow of the healing agent is restrained from filling the crack while when the value of Hf is low, the healing agent easily flows to fill the crack.

$$\text{Hf} = \frac{\text{viscous forces} * \text{contact area of liquid metal and solid}}{\text{volume expansion*density*gravity*liquid velocity*crack size}} \tag{1}$$

The simulations were validated by experiments. They noticed that a Hf larger than 407 should not be selected to observe complete healing of the crack. For values higher than 495, the liquid does not flow out of the container. However, the reaction between the healing agent and the matrix is not accounted for in the Hf parameter. Future work should be done to include these thermodynamics parameters.

2.1.2 Coating Agent

The second liquid-based self-healing mechanism consists of the use of a *coating agent* (see Fig. 1b). The general working principle of this healing strategy is similar to the previous system with the exception that the healing agent is not encapsulated in a tube or in a microcapsule but is deposited as a coating on the material to heal. Then, after cracking, a heat treatment triggers melting of the coating while keeping the rest of the material in solid state. The coating flows and fills the crack. The advantage of this strategy is that the composition and the microstructure of the base material is unchanged [17]. However, one major drawback is that the healing treatment has to be performed under protective atmosphere to avoid any oxidation, i.e. more complicated on site on large structures. In addition, the coating might affect the tribology behaviour of the material.

To the author's knowledge, only Leser et al. demonstrated the potential of this strategy [18]. A coating composed of 60 wt% of indium and 40 wt% of tin was deposited on Ti-62222 alloy by radio frequency deposition. As the melting temperature of the coating is ~124 °C, the healing treatment consists of a one-hour heat treatment at 135 °C. They observed that during fatigue crack growth testing at low crack-tip loads, the crack was arrested by this heat treatment. For higher crack-tip loads, the crack could not be arrested but a significant reduction in crack growth rate, typically 50%, was noticed. Moreover, they showed that this healing mechanism is repeatable, so at least two healing cycles may be applied on the same material. A healing experiment was performed in situ in the scanning electron microscope (SEM) after a scratch test on an Al coated with In-Sn, see Fig. 3 [18]. Some healing coating melted and flowed into the scratch but not in a homogeneous way. This is expectedly the consequence of surface oxidation. Indeed, this oxidised material shows a significantly higher melting temperature which prevents its flow inside the scratch and thus its healing.

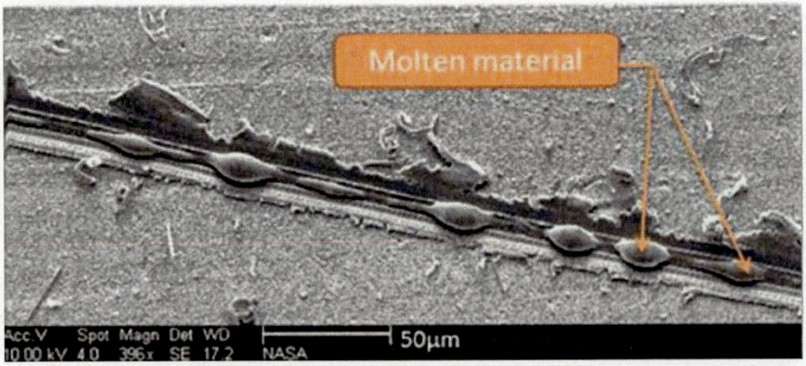

Fig. 3 SEM image of a healed scratch. Reproduced from [18]

2.1.3 Eutectic Healing

The third liquid-based healing strategy is called the *eutectic-based healing* (see Fig. 1c). In this mechanism, the microstructure of the alloy is composed of two phases: a high melting point dendritic phase and a low melting point eutectic phase [19]. It is this eutectic phase which will act as the healing agent. Indeed, after damage, a heat treatment in the liquid–solid temperature range will trigger the melting of the eutectic phase while the dendritic phase will maintain the structural integrity of the sample. This will allow the flow of the eutectic phase to the damaged sites and seal them during solidification.

In order to produce a eutectic-based healing system, an alloy composition away from the eutectic point should be selected. During solidification, dendrites of a saturated phase will first form while the composition of the remaining liquid changes to tend towards eutectic composition. At eutectic temperature, the remaining liquid will solidify into a eutectic phase in between the dendrites [19].

The main advantage of this method is that, in principle, many different systems can show this self-healing mechanism as long as it is hyper or hypoeutectic [19]. Moreover, the microstructure is quite simple (no reinforcement needed) and can be produced by classical manufacturing methods such as casting lowering the manufacturing costs.

However, designing such an alloy is not that simple, there are some challenges. First, a good ratio between dendritic and eutectic phase is required. Indeed, on one hand, enough dendritic phase is needed to maintain the structural integrity of the part. On the other hand, there must be enough eutectic phase for the liquid to reach the healing site and completely fill the crack [19]. Another issue of the eutectic-based healing strategy is that the healing agent is part of the material, i.e. the overall volume of the part does not change. Therefore, voids geometry is modified or moved to another location in the part where it would hopefully be less harmful [19]. In particular, crack tips, which are sources of stress concentration, are blunted by the liquid phase. In the systems of Sect. 2.2, this issue will be solved by using crack closure systems before crack sealing by the liquid.

Kim et al. have analysed this healing strategy in a damaged Al–Cu–Si–Sn–Bi alloy after warm rolling [20]. During a heat treatment at 150 °C (above the eutectic temperature), they observed that the Sn-Bi eutectic phase melted and filled the cracks; see Fig. 4.

For microelectronics applications, solder joints undergo thermomechanical fatigue which leads to high local stresses and cracking. To overcome this issue, research has been done on healing mechanisms in order to improve their fatigue life.

Siroky et al. developed a numerical model of the damage and healing behaviour of a binary Sn-Bi solder alloy [21]. The damage model was based on the concept of entropy developed by Basaran et al. [22] while the healing model was based on viscous material transport [23]. They showed that increasing the healing temperature and increasing the fraction of eutectic phase, increase the liquid fraction of healing agent and thus induce faster healing [21]. Increasing material mobility and the width of the liquid film also promote faster healing [24].

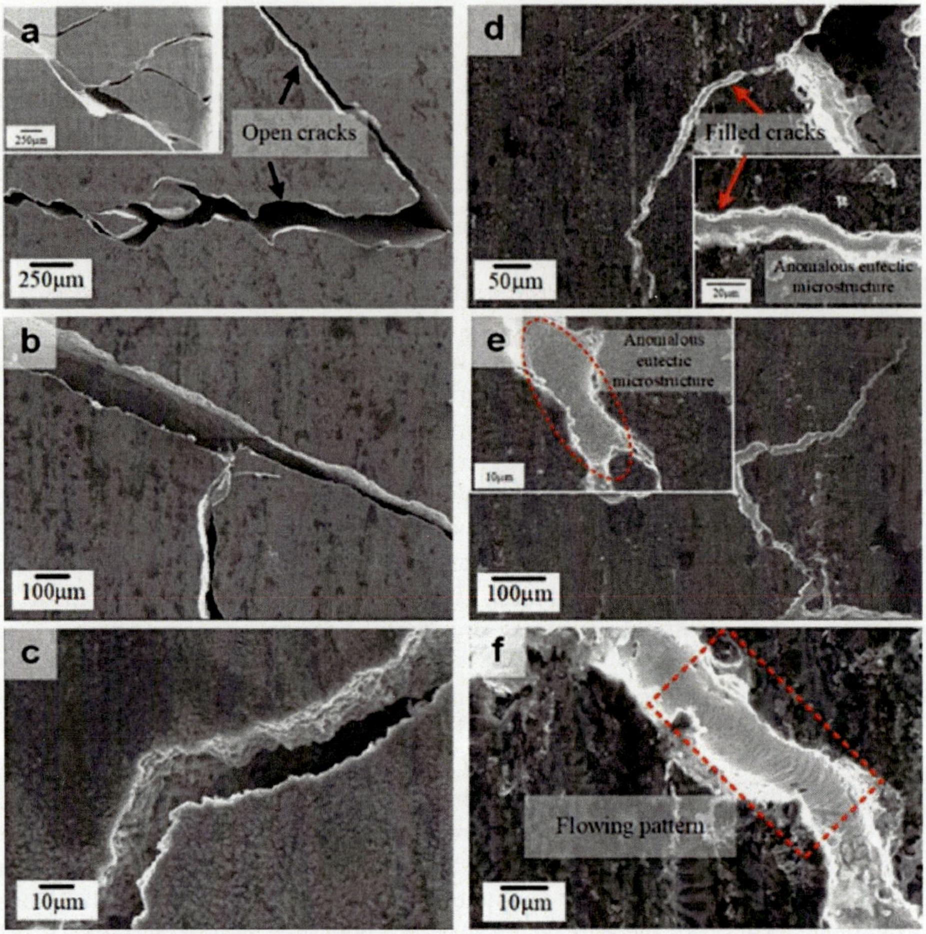

Fig. 4 SEM image of **a–c** damage in Al–Cu–Si–Sn–Bi alloy and **d–f** cracks filled after healing treatment. Reproduced from [20]

Subsequently, Siroky et al. analysed the healing ability of a Sn-40wt%Bi alloy. They experimentally observed in situ the healing following heating of a Sn-40 wt%Bi cube containing a spherical cavity made by indentation, as shown on Fig. 5 [25]. This cavity is completely filled after cooling. In a following publication, Siroky et al. [24] analysed the tensile properties of this Sn-40wt%Bi after healing. Tensile dog bone samples were loaded under cyclic tension until partial fracture, healed for 5 min at 140 °C and then reloaded again until fracture. They observed a healing effect and an effective restoration of the elastic mechanical properties. However, the fracture strain was significantly reduced from 28 to 10% and the ultimate tensile strength decreased by about 50%. SEM observations showed that this is due to partial healing caused by the incomplete wetting of the crack surface [24]. Therefore, self-wetting properties are essential parameters to analyse in the design of new healing materials.

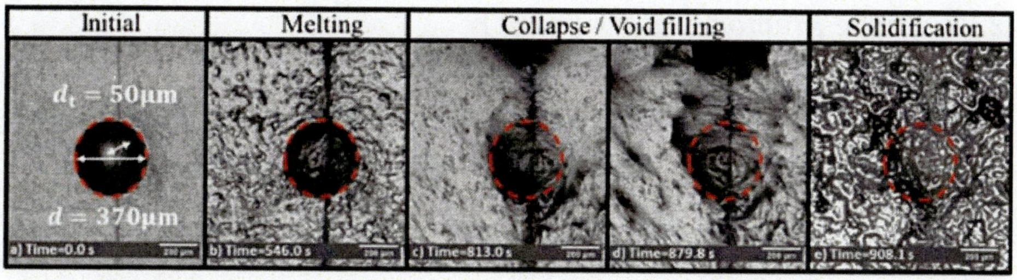

Fig. 5 Healing of a cavity in a Sn-40wt%Bi alloy during a heat treatment at 145 °C. Reproduced from [25]

Ruzek investigated two binary Sn-Bi systems: Sn-20wt%Bi and Sn-10wt%Bi which have respectively 21% and 29% of liquid phase at 170 °C. A crack was initiated using a bending test on a notched sample (see Fig. 6). He clearly observed by energy dispersive spectroscopy (EDX) some eutectic phase well bonded to the top surface of the notch and filling partially the notch [19]. Now, further research is clearly needed to completely fill cracks by the eutectic-based healing strategy.

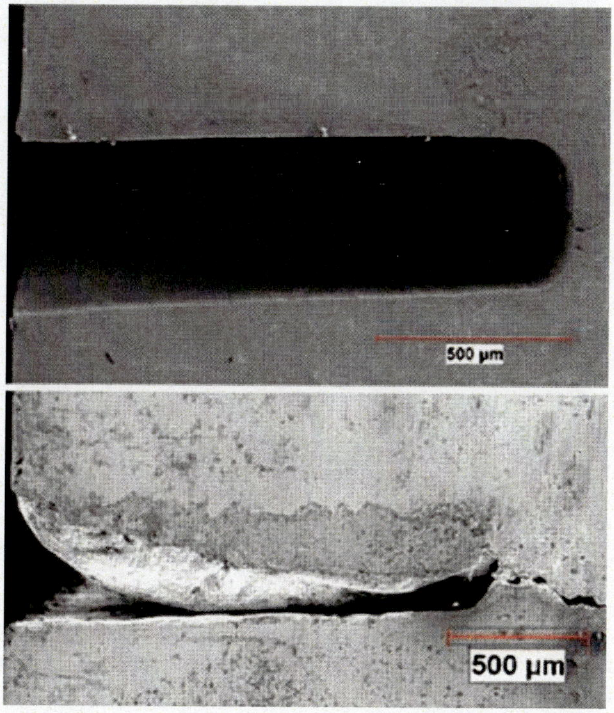

Fig. 6 SEM image of a crack in a Sn-10%Bi before and after healing treatment performed at 170 °C. Reproduced from [19]

2.2 *Shape Memory Alloys (SMA)*

Another macro-scale healing strategy reported in this chapter involves the use of shape memory alloys (SMA). After sustaining deformation, SMA are able to recover their initial un-deformed shape by a heat treatment. This treatment consists in heating above the phase transformation temperature of the SMA. Initially, the SMA is in a martensitic state but will transform under heating into its austenitic state (see Fig. 7) to recover its initial shape [17].

The most common associated healing strategy is called "**SMA-clamp and melt**" (see Fig. 1d). It focuses on healing macro-cracks. For this strategy, the material, which is essentially a composite, must contain two different constituents: SMA wires and a metallic matrix. Figure 7 details the different steps of this healing strategy. The SMA clamp and melt healing cycle can be divided in four steps [14]:

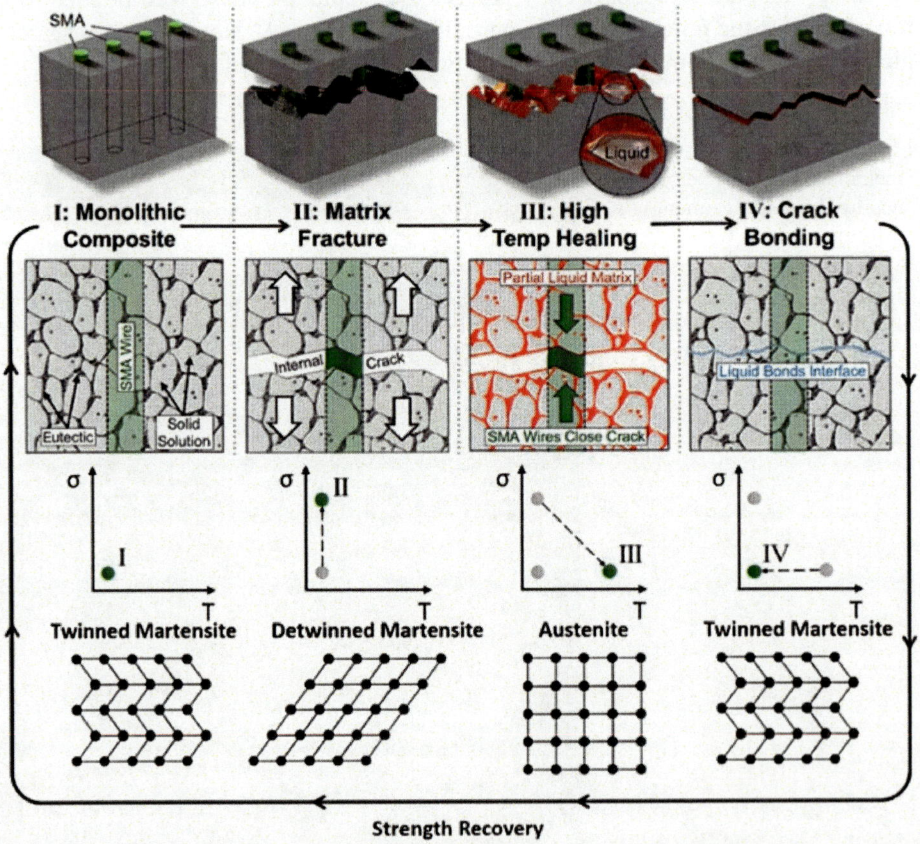

Fig. 7 Schematic representation of the cyclic healing of the SMA clamp and melt strategy. Stress and temperature state as well as phase transformations of the SMA wires are also schematised. Extended and inspired from [14]

- **Step I, initial state**: The metal matrix containing the SMA wires is at room temperature and is not exposed to any loading. The metal matrix is composed of two different phases similarly to the eutectic healing system of Sect.2.1.3: a dendritic and eutectic phase. They will be responsible for crack closing in step III. The SMA wires are in their twinned martensitic state. Usually, they are pre-strained before being incorporated to the matrix in order to exert a compressive force and close the crack in step III.
- **Step II, matrix fracture**: The system is exposed to loading in the wire direction. A crack will form only in the matrix if the stress level is larger than its ultimate tensile strength (UTS). However, this stress must remain below the UTS of the SMA wires to prevent their fracture. After this step, the SMA wires are still in their martensitic state but in the detwinned state.
- **Step III, high-temperature healing**: In this step, the system is not any more exposed to loading but a heat treatment should be applied. Its aim is twofold: trigger the SMA wires phase transformation and melting the eutectic phase of the metal matrix. If the phase transformation temperature of the SMA wires is reached, it will transform into austenite and recover its initial un-deformed shape. This phenomenon will put the two cracking surfaces in contact and eventually apply a compressive closing force if the wires were pre-strained. Concerning the melting of the eutectic phase, it will heal the crack according to same mechanism as in Sect. 2.1.3: it will allow the crack lips to be brazed together again.
- **Step IV, crack bonding**: In this final step, the system cools down to room temperature. As a result, the liquid eutectic solidifies and bonds the two cracking surfaces together, restoring structural integrity of the part. The SMA wires are also restoring their twinned martensitic state.

The main advantage of this strategy is that it could, in principle, be reproduced endlessly as the structure of the sample before and after healing is identical. However, it still is a non-autonomous strategy, and it is highly anisotropic. Indeed, the loading can only be applied in the wires direction otherwise the sample will experience critical failure [6].

This SMA clamp and melt strategy have been applied to various systems composed of a metal matrix and NiTi SMA wires. The first metal matrix to be investigated as a proof of concept of this strategy was Sn-13at%Bi combined with 1% volume fraction of NiTi wires. Manuel and Olson [26] used a healing temperature of 169 °C (leading to a 20% liquid fraction of the matrix). They recovered 95% of the UTS after healing. However, they also observed a significant reduction in ductility. Poormir et al. [27, 28] have taken the analysis of this specific system a step further by varying three different factors: the volume fraction of NiTi wires, their pre-strain level and the healing temperature. They observed a maximum UTS healing efficiency of 66% for 2.33% SMA volume fraction with 6% pre-strain and a healing at 190 °C, while the optimal ductility healing efficiency was observed for 0.78% SMA volume fraction with 0% pre-strain and a healing at 170 °C.

More interesting alloys from an application point of view were also examined in literature. Ferguson et al. [29] investigated the system containing Zinc ZA-8 matrix with 1% volume NiTi wires. They tested different load transfer strategies from the SMA wires to the Zn matrix, but they only managed to recover approximately 30% of the UTS after healing.

Al alloys are the most researched matrix material for this SMA clamp and melt strategy. NASA is interested in this healing strategy with the aim of healing fatigue cracks for aerospace aluminium parts. They obtained more than 90% recovery of UTS after healing an Al–Si alloy with 2% SMA wires [30]. Subsequently, Zhu et al. [31] identified with a finite elements simulation the optimal system parameters. They concluded that a higher healing temperature improves the softening of the metallic matrix improving the healing capability. Moreover, they demonstrated that pre-straining of the SMA wires is also beneficial.

Fisher et al. [14] paved the way to designing new aluminium system using this healing technique by fully characterising an Al–3Si with 2% volume of NiTi wires. They successfully obtained a 91.6% healing efficiency and showed promise for healing fatigue cracks while also exploiting the finite element simulation previously developed. All these works led to a patent filing by Manuel et al. [32].

Rohatgi [33] managed to reduce the width of a crack but did not achieved rebonding between fracture surfaces of a Al-A380 matrix reinforced with NiTi fibres. Others characterised more carefully the mechanical response of NiTi fibres [34] or investigated ways of detecting damage in SMA clamp and melt systems using ultrasonic structural health monitoring [35]. Similarly to the Sn-Bi study [26], Srivastava and Gupta [36] made a parametric experimental study on an AA2014 matrix with NiTi fibres varying the healing duration, the SMA wires volume fraction and diameter; see Fig. 8 for an example.

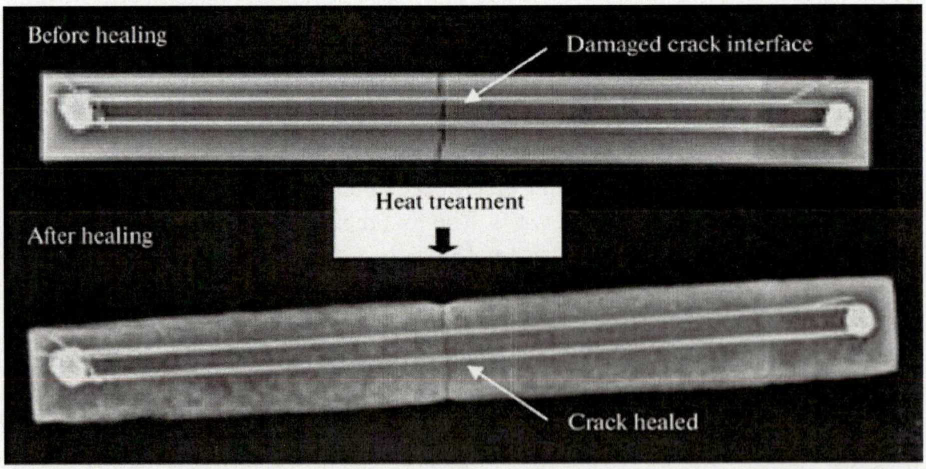

Fig. 8 X-ray radiography of an Al alloy AA2014 reinforced with NiTi wires before and after the healing treatment. Reproduced from [36]

2.3 *Electro-Healing*

Electro-healing process is represented in Fig. 1e, f. This strategy can be used to heal various size of defects. It differentiates itself from the preceding healing systems as here no heat input is required to trigger healing. Indeed, the healing process is driven by the application of an external electrical field [6, 37]. It is a particularly interesting strategy since it does not require any microstructure modification of the initial sample. This healing strategy can be divided into two categories: the first one will use an electrolyte solution as healing media (see Fig. 1e), while the second will use electropulsing (see Fig. 1f).

2.3.1 Electro-Healing Inside Electrolyte Solution

Zheng et al. [38] applied the first healing technique to Ni sheets. They managed to heal through-thickness cracks of about 1.5 mm; see Fig. 9. The samples were submitted to a current inside an electrolyte aqueous solution containing Ni ions for 2–20 h (depending on the sample thickness). During the electrical treatment, the ions were reduced at the surface of the crack and bonded with the Ni sheet. After healing, a similar yield and ultimate tensile strength to the base material were recovered for a 100-μm-thick sample. However, only a limited ductility was recovered. When thicker samples were healed, the efficiency of the healing process decreased. The thickness of the sample is thus a limiting factor for this healing strategy.

Hsain and Pikul [39] recently applied the same healing strategy to a cellular nickel structure coated with an insulating polymer; see Fig. 10. They were inspired by the healing process taking place in biological structural materials like bones. They transport nutrients and cells to the fracture site from other areas taking advantage of

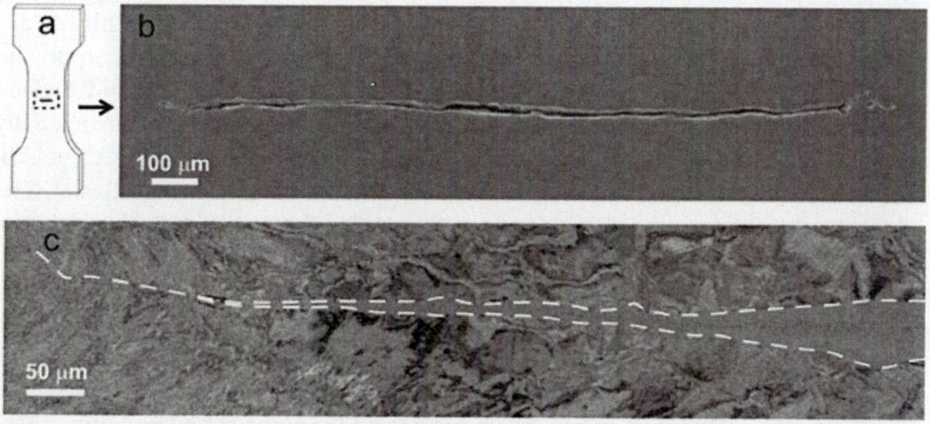

Fig. 9 **a** Representation of the through-thickness crack in a Ni sheet with its observation before healing in **b**. **c** SEM image of the healed crack after electro-healing treatment. Reproduced from [38]

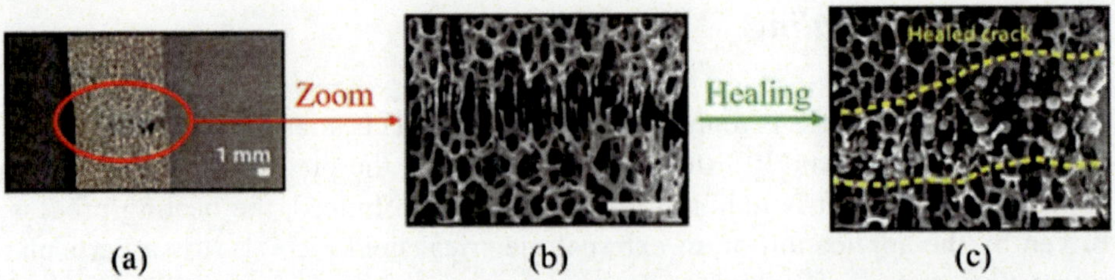

Fig. 10 **a** Photograph and **b** SEM micrograph of the cellular Ni structure before healing. **c** SEM micrograph of the healed cellular structure after application of an electrical field. Modified from [39]

their cellular structure to reach damage. In this synthetic healing system, the cellular Ni is plunged in an electrolyte solution to allow Ni ions to reach the fracture site. These ions repair damage by reduction when an external electric field is applied for 4 h. The insulating polymer coating prevents reduction taking place everywhere in the cellular Ni. After healing, the Ni network fully recovered its initial UTS.

2.3.2 Electro-Healing Without Electrolyte Solution

The second electro-healing system does not use an electrolyte solution to heal the sample, i.e. it is not an electro-chemical reduction reaction which is responsible for the healing process. For this healing technique, the application of an electropulsing treatment combined with the presence of defects like microcracks or microvoids will trigger the healing process. This treatment consists in applying a high-energy electric pulse whose duration is generally shorter than 1 ms. This makes electropulsing an extremely fast healing strategy. In order for a crack to be healed, two steps can be distinguished [40]:

- **Crack closure**: This is the first healing step. In order to successfully heal a crack, both cracked surfaces must be sealed together. To do so, electropulsing will generate thermal compressive stress around the crack or the void. Indeed, cracks and voids act as barriers to the electron flow, increasing resistance locally. Thus, an inhomogeneous thermal gradient is generated near the crack or void by the Joule effect which will generate a compressive stress locally [40–44]. This step corresponds to the shape memory effect used for the SMA healing strategy described in Sect. 2.2.
- **Crack healing**: The second healing step is the microwelding of the crack. Indeed, there must be metallurgical interactions between the two crack surfaces for them to be more than just sealed together. This healing can be explained by the local softening of the material near the crack tip generated by the local heating and compressive stress leading to a microwelding of the crack [41]. In addition, local heating of the material favours atomic interactions [40] and atomic diffusion [43].

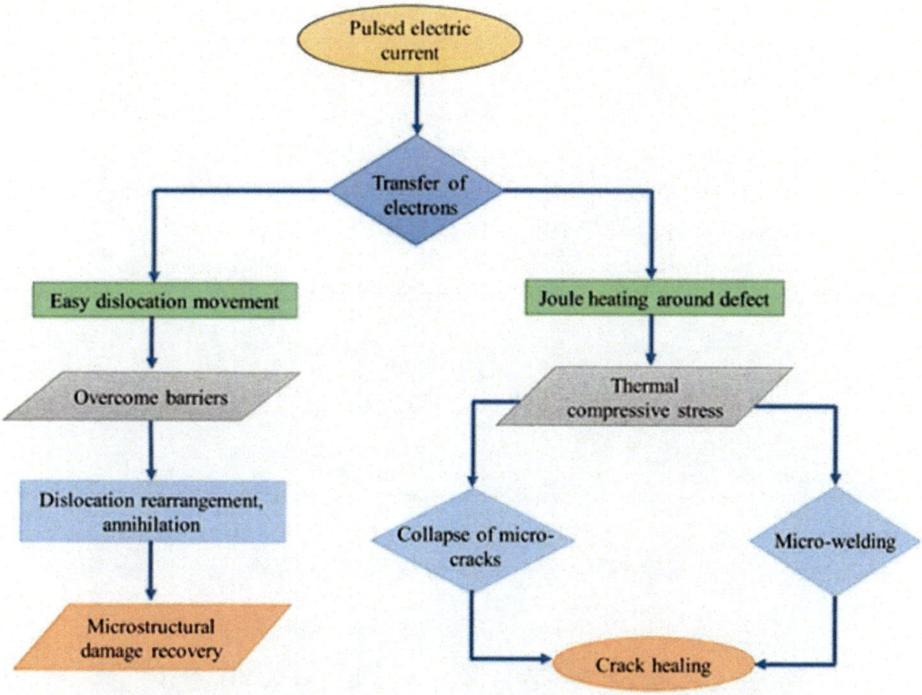

Fig. 11 Principle of crack healing using electropulsing. Reproduced from [41]

This step corresponds to the melting of a part of the microstructure used for the SMA healing strategy described in Sect. 2.2.

It was also reported that electropulsing increases dislocation mobility and reduces dislocation density. Meaning it may also recover or modify the material microstructure [41, 42, 45]. A schematic summarising this healing technique is provided in Fig. 11. The advantages of this healing strategy are that it works without any significant change to the material (no addition of SMA wires, coating or low melting point feature), and it also only affects locally the material on the damage site without having to locate it beforehand [40, 41].

The effect of electropulsing was investigated for several metals, steels being the most analysed. Zhou et al. [46] were the first to obtain a partial healing of a through-thickness crack for a medium carbon steel. A 350 μm crack was successfully healed. The same kind of crack healing was also observed on stainless steel by Yu and co-workers [47]. Tang et al. [45] focused on the microstructural effect of the electropulsing. They demonstrated that this electrical treatment delayed the fatigue crack initiation in a stainless steel due to a decreased dislocation density. Yang et al. as well as Wang et al. [43, 44] were interested in healing voids in TWIP steel and M50 bearing steel, respectively. They both showed a clear void healing effect of the electropulsing treatment while also showing a decrease of the amount of inclusions and carbides in both steels. Kumar and Paul [41] healed successfully an entire fatigue

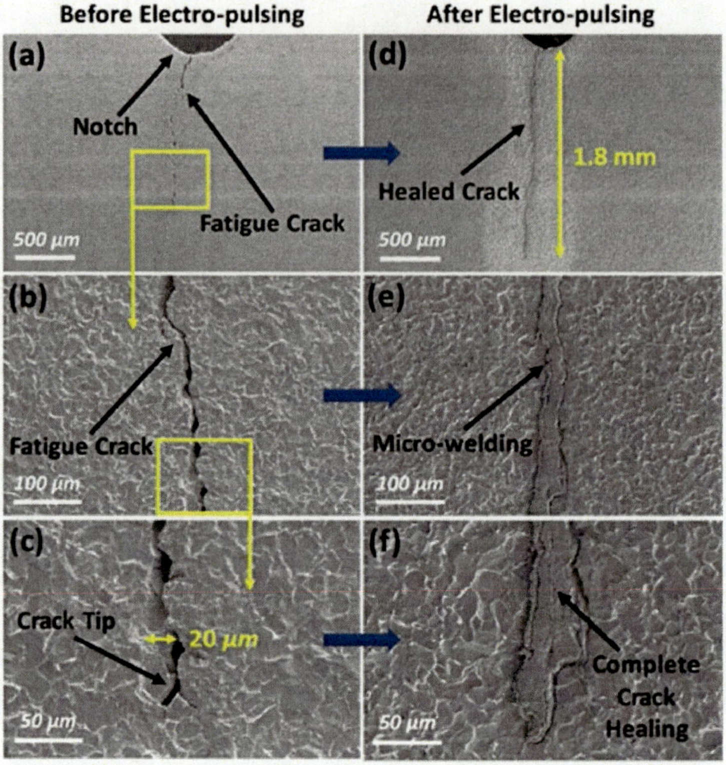

Fig. 12 SEM micrograph of a fatigue crack in a steel before and after electropulsing healing. Reproduced from [41]

crack of 1.8 mm in a drawn steel; see Fig. 12. Note that they used an electropulsing duration of 10 s, which is significantly longer than in the other references.

This electropulsing healing strategy has also been applied to two other metals: titanium and magnesium. Song and coworkers [40, 42] managed to heal microcracks in TC4 titanium sheet. They reached a recovery of more than 92% of the UTS of the base material and even improved ductility by at least 20%. They used finite elements simulations to show the non-uniform plastic deformation happening around the crack during electropulsing [48]. Finally, Xu et al. [49] investigated a magnesium alloy. They healed microcracks in an Mg tube by using a variant of electropulsing called “high-density eddy current pulse treatment”. As a result, the yield strength, the UTS as well as the elongation were improved compared to the initially damaged material (respectively a relative maximum increase of 24, 20 and 63%). They also identified that the squeezing action of the Lorentz force is responsible for crack healing.

Besides these structural applications, electro-healing is also used in the field of electronics for films or nano-wires. Only a few cases will be reported in this chapter since these are not construction materials. In electronics, the phenomenon called electromigration induces a migration of atoms when electronic components are subjected to high electric current density. This migration generates nano-voids in the components. The importance of this problem is increasing with the tendency to use smaller and more powerful devices. However, it is a reversible process that depends on the polarity of the current. Which means that damage generated by electromigration can be healed by inverting current polarity [50, 51]. As a result, it was shown that electromigration damage is a problem for DC current, while in AC current, the damage is healed by itself since the current periodically change its polarity [52]. So, electromigration could be used to heal nano-damage in small electronic components. Putz et al. [51] managed to heal nano-cracks using electromigration in a gold 50 nm thin film built on a flexible substrate. They applied a constant current density during 24 h to the sample. Baumans et al. [50] used electromigration to tune the resistance of superconducting nano-wires by controlling their void content. They managed to control it using either a feedback loop or by applying a precise number of electropulses (pulses of 1 s) to the wires.

Finally, Danzi et al. [53] performed a new type of crack healing that may be used to heal cracks in electronic films. Here, it is not the electromigration process that is used. They designed an on-chip healing system based on Ni/Al multilayers. The principle of their healing strategy is that these multilayers are characterised by a strong exothermic intermetallic-forming reaction that can be triggered by an electric pulse. This reaction generates a local temperature rise of up to 1500 °C that can weld locally a crack in an Au or Cu film; see Fig. 13. They managed to heal Au and Cu

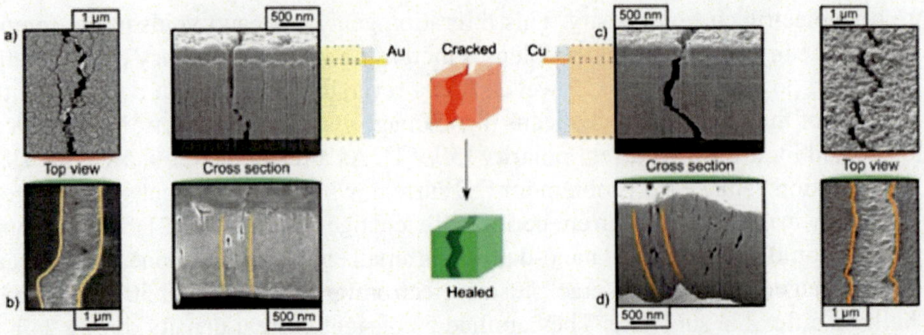

Fig. 13 **a** SEM micrographs of the Au thin film and Ni/Al multilayers before the healing and **b** SEM micrographs of the same system after the application of an electrical pulse that triggered the healing reaction. **c**, **d** SEM micrographs for the case of a Cu thin film. Reproduced from [53]

thin film cracks opened by up to 500 nm with the application of a current pulse but maintaining the system at room temperature. This healing strategy kept the healing time in the order of the millisecond.

3 Solid-State Healing of Nano- and Microscale Damage

Solid-state healing mechanisms are based on the activation of metallic atoms diffusion, i.e. activation of the healing agents mobility, in order to fill nano-scale voids and prevent macro-scale damage [13]; see Fig. 1g, h. Solid-state healing strategies require the availability and the sufficient mobility of healing atoms to move to the damage sites [6]. One of the main difficulties associated with metallic systems is the low mobility of the atoms at room temperature and the need for an external trigger to promote mass transfer without mechanical property drop [6, 10–12, 48].

It is thus important to understand the diffusional processes, and this section will start with a short discussion on solid-state diffusion in metals. Reviews on diffusion in metals can be found elsewhere [54]. Here, we focus on important aspects of diffusion in precipitation hardenable aluminium alloys and iron-based alloys, as these alloys have been the target of many studies. Furthermore, diffusion mechanisms are discussed depending on the external trigger used to promote diffusion: thermally activated diffusion or diffusion pipes linked to microstructural features, which can facilitate diffusion even at room temperature.

Solid-state healing strategies in metals will then be discussed and divided into categories depending on the nature of the healing agent (self-diffusion or solute atoms), the external trigger and the corresponding diffusion mechanism activated (thermal activation Fig. 1h or diffusion pipes at low temperature Fig. 1g). Furthermore, the more specific case of radiation-enhanced diffusion and corresponding healing strategy is also discussed. A short section summarises ongoing efforts regarding the modelling of solid-state healing strategies. Finally, an assessment of existing materials and further perspectives is discussed.

3.1 *Diffusion in Metals*

As already mentioned, the main difficulty associated with healing event in metallic systems is the low mobility of atoms at room temperature [10, 12].

Thermal activation is commonly used to promote diffusion of healing agents towards initiated defects. Thermally activated diffusion (TD) in metals is favoured at relatively high homologous temperatures and for relatively long time. These two characteristics are typical of creep applications and explains why important efforts have focused on the solid-state healing of creep damage in steels and aluminium alloys (see Sect. 3.4).

According to Van Dijk and co-authors [10], the applicability window for creep healing is limited between the lowest temperature at which creep deformation takes place ($T_{\min} = 0.4T_m$, with T_m the melting temperature) and the maximum temperature ($T_{\max} = 0.65T_m$) above which mechanical properties would become too low for practical applications. The evolution of the diffusion coefficient D with temperature is given by Arrhenius' equation, i.e. $D = D_0 \exp(-Q/RT)$ where D_0 is the frequency parameter [m^2/s], Q is the activation energy [J/mol], R is the gas constant, and T is the temperature [K] [10, 55–57]. The characteristic diffusion length d is then given by $d = 2\sqrt{Dt}$, where t is diffusion time in seconds. The characteristic diffusion length d gives an estimate of the size of the defect which can be successfully healed by diffusion.

Figure 14a,b show the self-healing potential, as estimated by the characteristic diffusion length, for an Al-based matrix and an Fe-based matrix. Figure 14a shows that, for solute Cu atoms acting as the healing agent in the Al matrix, the healable damage scale is of the order of 1 nm up to the order of a few μm at $T_{\min}$ and $T_{\max}$, respectively, and for an acceptable healing time (typically shorter than 24 h [10]). Similar results are observed for self-diffusion at $T_{\max}$. Figure 14b shows similar result, but for solute Au atoms acting as the healing agent in iron, with slightly higher healable damage ranging from a few *nm* up to a few dozen of μm at $T_{\min}$ and $T_{\max}$, respectively. In this case, the healable size is slightly decreased for the case of self-diffusion, i.e. indicating a slightly lower healing potential by self-diffusion. While this analysis only gives a first-order estimate, it appears clearly that the choice of the solute atoms which will serve as healing agents is a key parameter to design an effective self-healing metallic material.

Figure 14a,b also show that, at room temperature (i.e. without thermal activation), diffusion is very slow for transport of solute elements and the healable damage size becomes negligible (around 10^{-2} nm). However, non-thermally activated diffusion mechanisms might facilitate diffusion at room temperature. Indeed, most industrial metals have polycrystalline structures containing defects such as vacancies, dislocations, grain boundaries and surfaces. These defects act as diffusion "shortcuts", and each type of defect is associated with a type of diffusion [58] (Fig. 14c): vacancy diffusion (D_v), pipe diffusion along dislocations (D_p), grain boundary diffusion (D_{GB}) and surface diffusion (D_S).

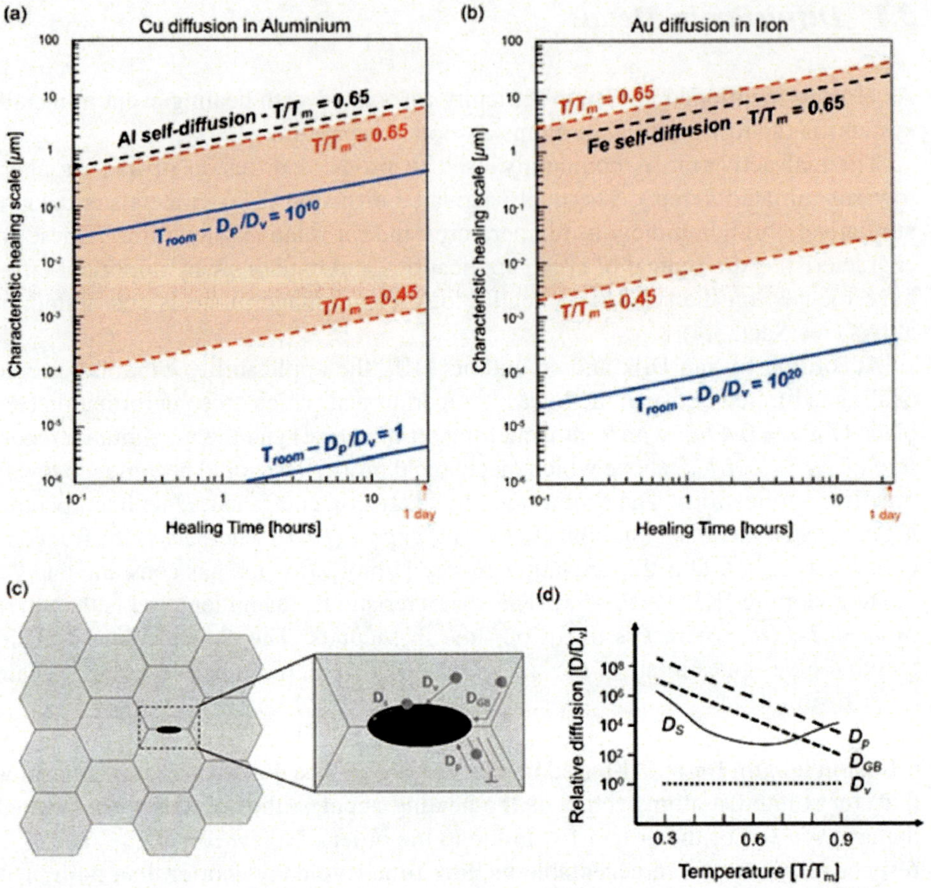

Fig. 14 Characteristic diffusion length as function of healing time **a** for self-diffusion and Cu in aluminium using $T_{\min} = 100$ °C, $T_{\max} = 330$ °C, Cu diffusion in Al: $D_0 = 6.5.10 - 5$ m^{2}s^{-1}, Q = 136 kJ.mol^{-1}[10], $D = 10$–15.7 m^{2}s^{-1} for self-diffusion at $T_{\max}$ [56], and **b** for self-diffusion and Au diffusion in Iron using $T_{\min} = 450$ °C, $T_{\max} = 900$ °C, Au diffusion in iron $D_0 = 7.10 - 5$ m^{2}s^{-1}, $Q = 227$ kJ.mol^{-1} [10] and $D = 10$–14.8 m^{2}s^{-1} for self-diffusion at $T_{\max}$ [57]. **c** Modes of atomic diffusion in polycrystalline metals (as defined in the text). **d** Typical variation of diffusion coefficients (normalised by D_v) as a function of the temperature in a polycrystalline metal (here Ag). **c**, **d** are inspired from [58]

The typical relative contributions compared to vacancy diffusion (D_v) of these different diffusion mechanisms and their temperature dependence are shown in Fig. 14d. It can be seen that generally D_{GB}, D_p and D_S become much larger than D_v near ambient temperature [58]. In order to achieve healing of a metallic system at room temperature, these diffusion shortcuts must be active within the damaged material [54]. Moreover, higher healing efficiency will be achieved by activation of all these diffusion mechanisms in parallel, especially at elevated temperatures.

If a metallic component is damaged, local plastic deformation surrounding the tip of a crack always generates regions with high dislocation densities [58]. Pipe diffusion along dislocations (D_p) can thus be considered as an intrinsic healing mechanism for most metals. This is also typical of long fatigue loadings during which cyclic deformations generate regions with high dislocation densities. Pipe diffusion has been widely exploited in Al alloys (see Sect. 3.3). Indeed, Fig. 14a shows that for D_p/D_V equals to 10^{10}, the healable damage size can reach almost 1 μm even at room temperature. However, for Fe-based systems, the healable damage size remains very small even for large values of D_p/D_V (such as 10^{20}; see Fig. 14b). This possibly explains the rather limited results for room-temperature healing for Fe-based systems [6].

To activate D_{GB}, the initial crack should be located in the vicinity of a grain boundary to provide access of healing atoms to the crack. Microstructures with finer grains size are thus likely to increase the healing efficiency but can be limited by technological aspects. Finally, surface diffusion (D_S) can also be considered as an intrinsic mechanism, since microcracks appearing during loading introduce free surfaces within any metallic system and facilitate diffusion.

This short discussion on diffusional processes in metals has identified key parameters to design an effective self-healing metallic system:

(1) The availability of solute atoms working as healing agents in solid solution, and their sufficient mobility is a key parameter.
(2) The applied heat treatment should be in a temperature window where only local mass transfer of mobile atoms occurs towards the crack plane and within a short period of time. This avoids critical microstructural changes and spontaneous precipitation of secondary phases not at the damage-induced defects which would lead to a reduction in healing potential.
(3) Controlled damage sites should preferentially be near the grain boundaries to exploit all diffusion mechanisms for healing.

The self-healing potential, as estimated by the characteristic diffusion length in simple binary systems (up to a few dozen microns for Fe-based systems at high temperature), gives interesting perspectives in terms of voids healing. However, the processes occurring in metals during heat treatments should be considered. In the case of Fe-based systems, for example, high healing temperature (such as 900 °C, i.e. T_{max}) for long healing time could induce phase transformations (such as austenitic transformation starting above 727 °C) and lead to a significant change in mechanical properties. Moreover, precipitation phenomenon taking place in most ternary systems can trap healing atoms into stable precipitates within the matrix, prohibiting them to reach the crack and reducing healing efficiency.

3.2 Healing by Self-Diffusion of the Matrix

Previous studies [11, 59, 60] have reported that the process of crack healing by self-diffusion may be divided into three stages (Fig. 15a):

- Blunting of the crack tips (stage *2* in Fig. 15b);
- Evolution of the crack into a channel of cylindrical voids and their continuous shrinking (stage *3* in Fig. 15b);
- Rapid final healing of the small remaining voids (stage *4* in Fig. 15b).

The healing kinetics depends strongly on several factors, such as crack location and geometry [59, 60], diffusion type [58, 59] and the concentration of solute atoms in the matrix [58].

To the best of the author's knowledge, several attempts were made to heal microcracks within Fe-based materials [60–62], but complete crack healing by self-diffusion was not reported. Cracks with sharp crack tips could be partially healed in pure α-Fe crystal; however, no further crack healing was observed [61]. Furthermore, healing by self-diffusion for Fe-based alloys presents several critical limitations. The main limitation remains the need of relatively high temperatures for healing activation which might lead to phase transformations [62] and impossibility to fully heal microcracks within a reasonable time period [62]. Moreover, free surfaces in metals are sensitive to oxidation which prevents the healing of open cracks [61].

Self-healing by self-diffusion in a sintered porous Ag deposited layer was observed by Chen et al. [63]. A partial crack closure on a notched tensile specimen was observed at 200 °C. TEM observations showed that the crack was healed by about 10-nm-size particles formed at the crack planes. As in the case of pure Fe, only the crack tip was healed after the heat treatment (even after 100 h).

Self-healing by self-diffusion was also observed for Au nano- and microparticles at 600 °C ($\approx 0.65T_m$ of Au, [64]). It was concluded that precisely tuning the annealing time, temperature and particle size leads to self-healing and full recovery of the initial shape of Au particles. Following nano-indentation, the healing mechanism has been associated with the accelerated self-diffusion along geometrically necessary dislocations forming terrace ledges.

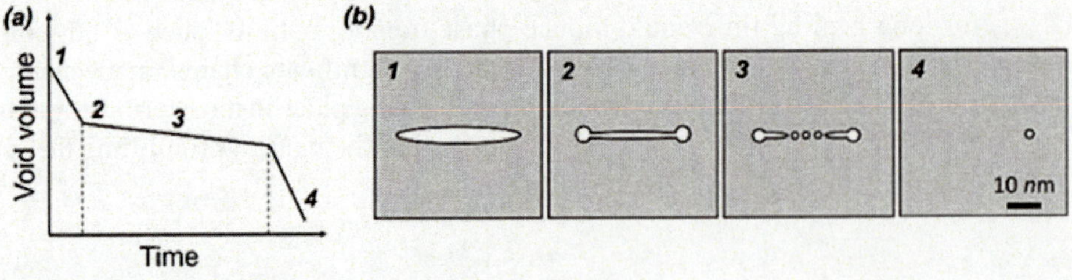

Fig. 15 Typical void healing by self-diffusion. **a** Evolution of the void volume as a function of time and **b** corresponding stages. Inspired from [11, 59]

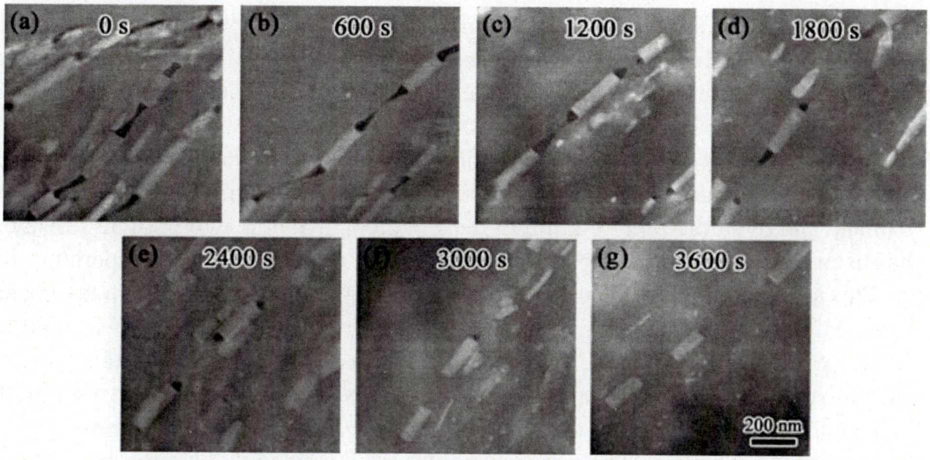

Fig. 16 STEM images of the annealed bulk Al–Mg–Er alloys after different annealing times. Reproduced from [11]

Finally, a successful example of self-healing of damage induced by cold rolling was demonstrated for an Al–Mg–Er alloy [11]. The damage related to the breakage of brittle inclusions was healed within 1 h at 200 °C (Fig. 8). TEM observation combined with EDX analysis revealed that healing happens by Mg atoms coming first to the crack planes and further Al atoms joining. This effect is due to the faster diffusion of Mg atoms in Al matrix than self-diffusion. After some heat treatment time, Mg is redistributed homogenously (Fig. 16).

3.3 *Healing by Dynamic Precipitation at Room T*

The availability and the mobility of solute atoms are essential for self-healing by dynamic precipitation at room temperature. Most research has thus been focused on precipitation hardenable aluminium alloys in underaged (UA) state. Indeed, the UA condition is characterised by alloying elements in supersaturated solid solution and provides a good healing potential. Furthermore, the other key element for effective healing at room temperature is the activation of diffusion shortcuts.

Hautakangas et al. [5] have studied Al–Cu–Mg alloy in underaged (UA) condition and during natural ageing. Using positron annihilation spectroscopy (PAS), they identified clustering of retained solute Cu atoms at deformation defects (after straining). In addition, the presence of dislocations and vacancies generated by straining facilitates diffusion, as discussed in Sect. 3.1. Furthermore, Nagai et al. [65] have shown that quenched-in vacancies in Al–Cu-Mg alloy are bound to Mg atoms. These atoms migrate along the dislocations and form vacancy-Mg-Cu complexes during early stages of ageing at 150 °C. This mechanism facilitates healing in that alloy.

Self-healing by dynamic precipitation has mainly been applied to enhance fatigue life performance. Indeed, during fatigue loading, microplasticity generates regions of substantial vacancies and high dislocation densities. Again, these defects act as diffusion shortcuts and facilitate diffusion. Fatigue damage is also more likely to nucleate in regions of stress concentration due to dislocations pile-ups [13], which subsequently favours dynamic precipitation by pipe diffusion. The concomitance of damage and dynamic precipitation in these regions explains why fatigue resistance has been the key property targeted by self-healing of metals at low temperature [6].

This idea was applied to 2xxx series Al alloys (Al–Cu–Mg based) also to enhance fatigue life performance in underaged (UA) state. Improved fatigue life was reported for 2xxx series alloys in the UA condition compared to the peak-aged state [66–69]. Recently, Zhang et al. [70] have proposed a very similar concept for exploiting the mechanical energy imparted during the initial cycles of fatigue to improve fatigue strength of UA Al alloys. This improvement is associated with dynamic precipitation in precipitate-free zones (PFZs) adjacent to grain boundaries. Even if it is not self-healing per se (but rather training of the material) and can only be applied to Al alloys containing PFZs, fatigue life of high-strength Al alloys was improved by a factor 25.

While improved fatigue resistance has been observed in commercial Al–Cu–Mg alloys [66, 68], confirming self-healing capacity, limited attention has been given to this healing strategy due to several issues [6]. The nature of fatigue damage [71] in real applications might explain the limited interest in this approach. Indeed, Wanhill [72] has shown that, in Al alloys, fatigue damage involves almost always defects in near-surface regions where environmental effects could potentially hinder crack healing [6, 72]. Furthermore, healing by dynamic precipitation is restricted to primary nano-scale damage [6, 10] at initial stages before extended fatigue crack growth. Thus, this mechanism might not be effective for repairing larger fatigue damage related for example to the initial porosity or constituent particles which are always present in commercial Al alloys [71, 73]. Moreover, the contribution of self-healing to the fatigue life improvement remains unclear and cannot be separated from the effect of dislocation motion pinning by dynamic precipitation and solute atmospheres [74]. Thus, the unambiguous link between healing by dynamic precipitation and fatigue life improvement has not yet been established [6, 69].

Even disregarding all surface-related challenges, the concept of self-healing strategy for fatigue life improvement is considered as inapplicable in practice for industrial applications [72] and further development is needed. Indeed, the main limitation remains the healable damage size which should be at least above 10 μm, while current approaches to healing by dynamic precipitation apply only to damage size below about 1 μm [72]. Thus, existing self-healing solid-state strategies cannot be considered as a solution to overcome the general need in damage tolerance improvement of aluminium parts. However, if further improved, they could contribute to improving the fatigue lifetime and general cost-effectiveness.

3.4 *Self-Healing During Creep*

As already discussed, relatively high homologous temperatures and relatively long service time facilitate solid-state diffusion in metals, and thus self-healing of creep damage has attracted significant research attention [6]. Furthermore, creep of metals usually involves cavities formed at the grain boundaries, ensuring grain boundary diffusion provides access of healing agent to the damage sites.

3.4.1 Fe-Based Systems

He et al. [75, 76] have shown that Fe-based alloys demonstrate increased precipitation of Cu on defects induced by plastic deformation in as-quenched samples compared to annealed ones [76]. A pre-strain favours these phenomena [75]. This improved diffusion and precipitation of Cu might appear similar to what has been discussed previously in Sect. 2.3 for Al alloys at low temperature [5]. However, the temperatures are very different: while for Al alloys, precipitation was detected during natural ageing (i.e. at room temperature), ageing was performed at elevated temperature (550 °C) in the case of Fe-based alloys.

This effect of Cu precipitation during creep test was demonstrated by Laha et al. [77, 78] for austenitic stainless steel 347 with microalloying of Cu, B and Ce. On the one hand, the addition and precipitation of Cu solute atoms decrease cavity growth rate. On the other hand, the addition of Ce prevents S segregation and favours instead B segregation on cavities surface. Due to higher melting point of B, the surface diffusivity is reduced retarding further cavity growth and improving creep performance of the steel. Concomitantly, Shinya et al. [79] have shown that compositional modification of austenitic stainless steel by adding Ti, B, N and Ce can prevent S segregation on creep cavities as well and instead provide BN films precipitation on the surfaces of creep cavities increasing the creep rupture strength and ductility of the steel (Fig. 17).

Other candidates for pure Fe, e.g. Au [37], were considered as healing agents by researchers from TU Delft. Although Au does not seem the right solution due to its cost, Au atoms are good healing agent candidates since they present all features of ideal alloying elements [37]. First, Au atoms remain in solid solution due to a high-energy barrier preventing homogeneous nucleation. This effect is linked to their larger size resulting in high strain energy. Second, presence of dislocations

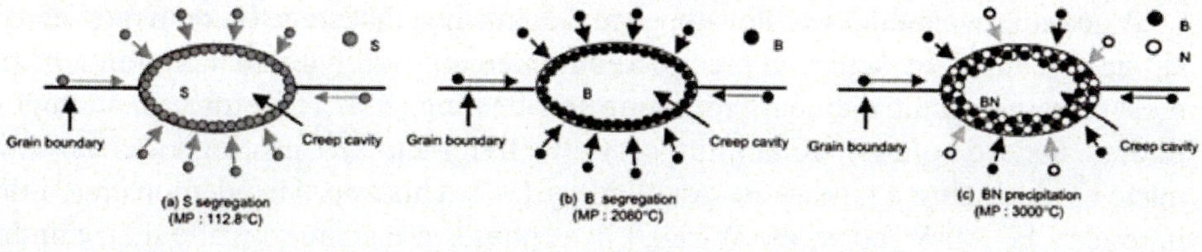

Fig. 17 An illustration of S and B segregation and BN precipitation on creep cavity surface. Reproduced from [79]

induced by straining can reduce this energy barrier by heterogeneous nucleation of Au precipitates. These two factors help to achieve selective precipitation at the damaged sites, preventing precipitation throughout the matrix, thus overcoming one of the main difficulties, i.e. the loss of beneficial effect of healing atoms being trapped in stable precipitates. Indeed, experimental results have shown that Au precipitates were detected only in pre-deformed samples, meaning that Au precipitation occurs solely on defect [37]. Moreover, in Fe–Au–B–N alloy, B and N presence overall retards Au precipitation. In contrast, in Fe–Cu–B–N alloy, B and N addition did not retarded overall Cu precipitation, but promoted formation of spherical precipitates throughout the matrix [75].

Experimental results of this approach were reported by the same group of authors. Au precipitation was clearly observed in Fe-Au [37] and Fe–Au–B–N [80] alloys in cavities located at grain boundaries. Au precipitation was proved to increase the creep performance compared to a Au-free Fe–Cu alloy [81]. Improved creep properties, related to the filling of nano-scale cavities by Au diffusion, were also facilitated by grain boundary and pipe diffusion [82]. It was shown that optimal temperature to achieve Au atoms mobility is around 550 °C. At a stress level of 90 MPa, a maximum filling ratio of about 80% was achieved, i.e. most cavities were filled completely. Further X-ray nano-tomography confirmed precipitates nucleation at grain boundaries near open-volume defects. In addition, a strong correlation between the cavities and precipitates morphologies is reported [83].

Since gold has limited practical applications, density functional theory (DFT) modelling of diffusion through a single-vacancy mechanism was performed for 28 elements in order to extend this approach to other binary systems [84]. In line with this effort, classical alloying elements for creep steels were chosen for further research, such as Mo [85] or W [86], in order to make a step towards industrial application.

Experimental analysis performed on Fe-Mo (6.2 wt% Mo) binary alloy have shown that creep cavities were partially filled with so-called Laves phase Fe_2Mo; see Fig. 18 [85]. Nevertheless, Fe-Mo system was found to be a less effective healable alloy compared to Fe-Au [12]. Further attempt with high-purity Fe-W alloy [86] demonstrated the same phenomenon of selective Laves Fe_2W precipitates. With the support of 3D nano-imaging, it was shown that the precipitates are also adapting themselves to cavity shape. Besides, isolated cavities tend to be fully filled, while interconnected cavities continue to grow. In conclusion, since both Mo and W are interesting in steels for solid-solution strengthening, a precise design of composition and their additional supersaturation can provide incorporation of this effect with desirable self-healing property and additionally extended creep life.

A quantitative model was developed and confirmed that stage II creep rates can be strongly reduced by selective precipitation on creep cavity and that Au remains the most effective solute element promoting self-healing [87]. Therefore, an attempt to find a balanced solution combining cost-effectivity and self-healing efficiency was made by designing a ternary Fe-Au-W alloy [88]. This approach demonstrated that in ternary Fe-Au-W alloy, the W-rich Laves phases are smaller, while their number density is increased compared to the Fe-W binary alloy. Concomitantly, Au-rich precipitates are not different in terms of size and precipitation kinetics for Fe-Au-W

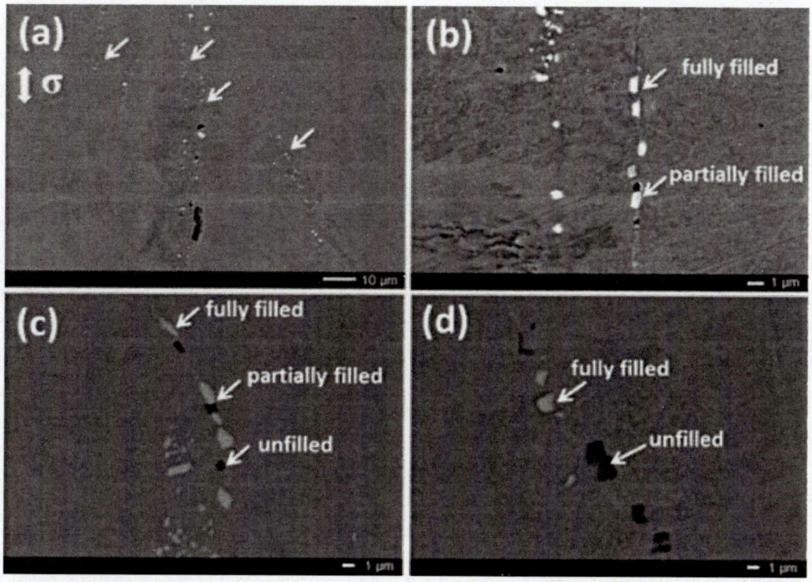

Fig. 18 Micrographs of the Fe-Mo alloy after creep for a stress of 160 MPa at a temperature of 838 K (565 °C) demonstrating cavities and precipitation at grain boundaries parallel to the loading direction at selected locations (*a* to *d*). Reproduced from [85]

and Fe-Au alloys. The addition of Au is thus enhancing the self-healing potential of Fe-W alloys, as confirmed for a Fe-3Au-4 W (wt%) alloy [89]. Indeed, this alloy has shown the longest creep life compared to previously studied Fe-Au [82] and Fe-W [86] alloys. Au- and W-rich precipitates coexisted on creep-induced cavities, and the filling ratio achieved about 70% for the lowest applied stress. Interestingly, Au-rich precipitates are micron-sized and able to fully fill the induced cavities, while W-rich precipitation are nano-sized, thus not able to fully fill cavities. Moreover, Au-rich precipitates are formed at the earlier creep stages by grain boundary and bulk diffusion, while W-rich precipitation is postponed to later creep states and controlled by solute diffusion. Overall, all precipitates are generally finer, and their number density is higher than that in corresponding binary alloys, which can additionally contribute to strength.

3.4.2 Al-Based Systems

For 2xxx series aluminium alloys [90, 91], the advantages of secondary precipitation by pipe diffusion were demonstrated also for creep applications. Improved creep performance is related to dynamic precipitation of S (Al_2CuMg) and θ' phases at grain boundaries. Moreover, Lumley et al. [92] showed that the creep resistance of an Al–Cu–Mg–Ag alloy is significantly increased for UA condition (compared to peak-aged condition) as healing atoms are available in solid solution for dynamic precipitation.

Creep is easier to heal in contrast to fatigue defects in near-surface regions where environmental effects could potentially hinder crack healing [6, 72]. However, one major challenge associated with healing of creep damage in Al alloys is that precipitation should not happen spontaneously throughout the microstructure, i.e. it should only happen on damage sites. Indeed, Lumley et al. [92] showed that increasing the time at room temperature between ageing and creep testing progressively reduces the beneficial effect of the UA treatment due to a secondary precipitation throughout the matrix. Nevertheless, the proposed strategy can be applied to other age-hardenable Al alloys [92] and thus still offers interesting perspectives for future research. Furthermore, the clear evidence of direct effect of the early stage damage healing to the creep performance was not yet reported [10] and remains an open question.

3.5 Self-Healing of Radiation Damage

Self-healing of radiation-induced damage has recently received major attention [12]. Some predictive models of self-healing behaviour were proposed for Fe [93], Cu [94] and W [95], where self-healing of radiation damage was attributed to the phenomenon of interstitials and vacancies' annihilation near grain boundaries. The effect of full vacancies annihilation was demonstrated for amorphous ZrCu alloys [96, 97] in conjunction with a good resistance to neutron irradiation. Moreover, it was shown that self-healing response depends on the microstructure and alloys composition.

Based on self-healing of creep damage in high-purity Fe-based systems, Zhang et al. [98, 99] transferred the binary Fe-Au and Fe-Cu systems to radiation damage. These studies demonstrated a similar precipitation behaviour compared to creep loading for Fe-Au alloy, while no significant precipitation was observed for Fe-Cu alloy. Indeed, Fe-Au alloy has demonstrated improved radiation resistance and irradiation hardening effect probably due to preferential Au precipitation at induced damage sites.

3.6 Modelling

Molecular dynamics (MD) models were developed in order to predict healing ability for Fe [100], Al [101] and Cu [102, 103] crystals. Figure 19 shows crack healing predictions for Cu nano-plate at 750 K [103]. Healing starts from crack tips blunting with further shape rounding accompanied with slip band movement until the crack fully disappears.

These modelling results cannot be directly reproduced experimentally due to the difficulty to match the exact conditions and the limited interest in use of pure metals. However, summarising the results for all three pure metals allowed to confirm two main key concepts already mentioned above: (i) the thermal energy is the driving force

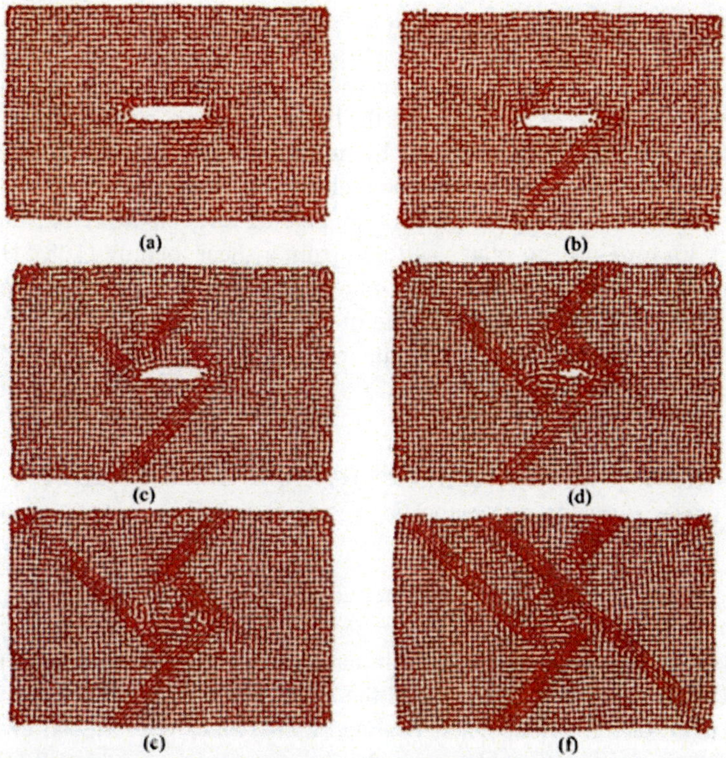

Fig. 19 Modelling results for the crack healing process at the temperature of 750 K: **a** 3.5 ps, **b** 5 ps, **c** 6 ps, **d** 7 ps, **e** 7.2 ps, **f** 18 ps. Reproduced from [103]

for crack healing and (ii) the presence of defects around crack tips will significantly influence healing kinetics [100–102].

Aside of healing process modelling by self-diffusion, coupling a kinetic Monte Carlo technique and a standard finite element was exploited by Karpov et al. [104]. They investigated macroscopic static loading and cavity geometry effects on the healing by interstitial diffusion and surface precipitation in an austenite nano-crystal. Cavities of elliptic cross sections were considered. The results have shown that healing time decreases with the applied loading magnitude and depends directly on the cavity volume. However, the clear aspect ratio effect was not evidenced.

Additionally, finite element modelling (FEM) was applied to analyse self-healing behaviour in creep steels [105]. It was shown that the filling of open-volume creep cavities located at grain boundaries is triggered by solute atoms flux along grain boundaries or from the bulk. The time required to fill the void depends on D_v and D_{GB} parameters, and the type of activated diffusion route will depend on their ratio. Another important parameter influencing the self-healing is the intercavity spacing with respect to the creep cavity size.

General rise of interest in self-healing metallic materials and recent experimental success provoked a huge interest in self-healing behaviour modelling and prediction. Recently, several new mechanisms were proposed which are not yet demonstrated experimentally. Among them is the healing of nano-cracks in nano-crystalline nickel by generation of disclinations by migrating grain boundaries [106] at room temperature or by stress-driven grain boundary migration during creep [107]. Another concept was shown for a room-temperature healing through the dislocation emission at a crack tip for Cu nano-plates suffered from a shear loading [108]. However, all these suggestions are idealised cases, pure metals and nano-scale damage; predictive models can promote understanding of self-healing behaviour of already proven concepts as well and help to improve their performance.

3.7 *Summary and Perspectives for Solid-State Healing Systems*

Solid-state self-healing strategies related to creep applications received much attention during the last decade. A summary of the solid-state healing systems detailed in this section is provided in Table 1. Summary of solid-state healing systems. Several successful results were achieved, and the authors believe that further investigation can lead to the development of self-healing Fe-based alloys for industrial creep application in the coming years. However, the same result is not achieved for Al alloys. New strategies need to be developed for this class of materials, and several attempts were indeed proposed, such as that of Michalcova et al. [109] for Al-30Ag system, but the considered high-purity material and applied heat treatment of 550 °C have additional limitations. Nevertheless, to develop an applicable self-healing strategy by solid-state diffusion, several limitations need be to overcome:

- Precipitation should happen solely on the damaged sites, avoiding trapping of healing solutes in the stable secondary phases throughout the matrix.
- Healing should be effective for both nano- and microvoids.
- Working temperature should be scrupulously controlled in order to limit microstructural changes and reduced mechanical properties.
- Microalloying selection and working conditions necessary for self-healing should be cost-effective.

The discussed strategies were mainly describing Al- and Fe-based alloys, giving a huge perspective to investigate self-healing mechanisms in other metals and alloys. As an example, Ni- or Ti-based alloys could be considered as potential systems with self-healing ability especially at elevated temperatures. Moreover, further development of self-healing for pure metallic nano-crystals can be of importance for electronic applications.

Table 1 Summary of solid-state healing systems

Material	Damage type	Healing agent	Healed size	Healing type	References
Al–Cu–Mg–Ag	Creep	Cu and Mg atoms	Not evident, nano-range	Autonomic/Intrinsic	[90, 92, 110]
Al–Cu–Mg–Ag	Fatigue	Cu and Mg atoms	Not evident, nano-range	Autonomic/Intrinsic	[66, 67]
Al–Cu–Mg	Uniaxial straining	Cu and Mg atoms	Not evident, nano-range	Autonomic/Intrinsic	[5, 65]
Al–Mg–Er	Cold rolling	Mg atoms	Nano-range	No autonomic/Intrinsic	[11]
Fe–Cu–B–N	Uniaxial straining	Cu, B, N atoms	Not evident, nano-range	No autonomic/Intrinsic	[75–79]
Fe-based (steel 1045)	Plate impact technology	Matrix self-diffusion	Microrange	No autonomic/Intrinsic	[62]
Pure Fe	Fatigue	Matrix self-diffusion	Microrange	No autonomic/Intrinsic	[60]
Pure Fe	Creep	Au, Mo, W precipitates	Nano- and microranges	Autonomic/Intrinsic	[37, 80–83, 85, 86, 88, 89, 111]
Pure Fe	Irradiation	Au	Nano	Autonomic/Intrinsic	[98, 99]

4 Conclusions

Up to now, all self-healing strategies are commonly divided into two size classes: nano- or macro-scale damage healing strategies [6]. The only actually really autonomous self-healing mechanisms (i.e. at room or under service temperature) involve solid-state diffusion. Their drawback: they can only heal nano-scale damage [6]. Current macro-scale healing strategies are non-autonomous, involving a heat or electrical treatment. [6]. They however have the big advantage to have the potential to totally heal highly damaged materials or fatigue cracks.

Acknowledgements This chapter writing has been funded by the European Research Council (ERC) under the European Union's Horizon 2020 research and innovation program (grant agreement n°716678). M.A., J.G and N.N. acknowledge the support of the Fonds de la recherche scientifique—FNRS (FRIA grant), Belgium. F.H. acknowledges the FNRS for his postdoctoral fellowship at UCLouvain.

References

1. Raabe D, Tasan CC, Olivetti EA (2019) Strategies for improving the sustainability of structural metals. Nature 575(7781):64–74
2. van der Zwaag S, Brinkman E (2015) Self healing materials: pioneering research in the Netherlands. IOS Press.
3. White SR, Sottos NR et al (2001) Autonomic healing of polymer composites. Nature 409(6822):794–797
4. Bailey BM, Leterrier Y et al (2015) Electrically conductive self-healing polymer composite coatings. Prog Org Coat 85:189–198
5. Hautakangas S, Schut H et al (2008) Self-healing of deformation damage in underaged Al–Cu–Mg alloys. Scripta Mater 58(9):719–722
6. Grabowski B, Tasan C (2016) Self-healing metals. In Self-healing materials, pp 387–407
7. Sharma S, Nandan G et al (2019) Recent advances in self-healing materials. Mater Today Proc 18:4729–4737
8. Nosonovsky M, Rohatgi PK (2011) Development of metallic and metal matrix composite self-healing materials. In Biomimetics in materials science, pp 87–122
9. Srivastava V, Gupta M (2018) Approach to self healing in metal matrix composites: a review. Mater Today Pro 5(9, Part 3):19703–19713
10. van Dijk N, van der Zwaag S (2018) Self-healing phenomena in metals. Adv Mater Interfaces 5(17)
11. Song M, Du K et al (2014) In situ electron microscopy investigation of void healing in an Al–Mg–Er alloy at a low temperature. Acta Mater 69:236–245
12. Zhang S, van Dijk NH, van der Zwaag S (2020) A review of self-healing metals: fundamentals, design principles and performance. Acta Metallurgica Sinica (English Letters) 33(9):1167–1179
13. Hager MD, Greil P et al (2010) Self-healing materials. Adv Mater 22(47):5424–5430
14. Fisher CR, Henderson HB et al (2018) Repairing large cracks and reversing fatigue damage in structural metals. Appl Mater Today 13:64–68
15. Martínez Lucci J, Amano RS, Rohatgi PK (2016) Heat transfer and fluid flow analysis of self-healing in metallic materials. Heat Mass Transfer 53(3):825–848

16. Martinez Lucci J, Amano RS et al (2008) Experiment and computational analysis of self-healing in an aluminum alloy. In ASME 2008 international mechanical engineering congress and exposition
17. Kilicli V, Yan X et al (2018) Recent advancements in self-healing metallic materials and self-healing metal matrix composites. Jom 70(6):846–854
18. Leser P (2014) Mitigation of crack damage in metallic materials. National Aeronautics and Space Administration, Langley Research Center
19. Nosonovsky M, Rohatgi PK (2011) Biomimetics in materials science: self-healing, self-lubricating, and self-cleaning materials. Springer, New York
20. Kim J, Kim HJ et al (2018) Thermally-triggered dual in-situ self-healing metallic materials. Sci Rep 8(1):2120–2120
21. Siroky G, Kraker E et al (2019) Numerical study on local effects of composition and geometry in self-healing solders. In 2019 20th international conference on thermal, mechanical and multi-physics simulation and experiments in microelectronics and microsystems (EuroSimE)
22. Basaran C, Yan C-Y (1998) A Thermodynamic framework for damage mechanics of solder joints. J Electron Packag 120(4):379–384
23. Liu P, Yao P, Liu J (2008) Effect of SiC nanoparticle additions on microstructure and microhardness of Sn-Ag-Cu solder alloy. J Electron Mater 37(6):874–879
24. Siroky G, Kraker E et al (2021) Effect of solder joint size and composition on liquid-assisted healing. Microelectron Reliab 119:114066
25. Siroky G, Melinc D et al (2020) Healing solders: a numerical investigation of damage-healing experiments. In 2020 21st International conference on thermal, mechanical and multi-physics simulation and experiments in microelectronics and microsystems (EuroSimE)
26. Manuel MV, Olson G (2007) Biomimetic self-healing metals. In 1st International conference on self-healing materials
27. Poormir MA, Khalili SMR, Eslami-Farsani R (2018) Optimal design of a bio-inspired self-healing metal matrix composite reinforced with NiTi shape memory alloy strip. J Intell Mater Syst Struct
28. Poormir MA, Khalili SMR, Eslami-Farsani R (2018) Investigation of the self-healing behavior of Sn-Bi metal matrix composite reinforced with NiTi shape memory alloy strips under flexural loading. Jom 70(6):806–810
29. Ferguson JB, Schultz BF, Rohatgi PK (2015) Zinc alloy ZA-8/shape memory alloy self-healing metal matrix composite. Mater Sci Eng, A 620:85–88
30. Wright MC, Manuel M, Wallace T (2013) Fatigue resistance of liquid-assisted self-repairing aluminum alloys reinforced with shape memory alloys shape memory alloy self-healing (SMASH) technology for aeronautical applications. National Aeronautics and Space Administration
31. Zhu P, Cui Z et al (2016) Characterization and modeling of three-dimensional self-healing shape memory alloy-reinforced metal-matrix composites. Mech Mater 103:1–10
32. Manuel MV, Fisher CR, Wright MC (2020) Self-repairing metal alloy matrix composites, methods of manufacture and use thereof and articles comprising the same. University of Florida Research Foundation Inc., United States of America
33. Rohatgi PK (2014) Al-shape memory alloy self-healing metal matrix composite. Mater Sci Eng, A 619:73–76
34. Salowitz N, Correa A et al (2018) Mechanics of nickel–titanium shape memory alloys undergoing partially constrained recovery for self-healing materials. J Intell Mater Syst Struct
35. Suri S, Correa A et al (2016) Initial integration of ultrasonic stuctural health monitoring of self-healing materials. In 8th European workshop on structural health monitoring. Bilbao, Spain
36. Srivastava V, Gupta M (2020) Parametric assessments of self-healing characteristics in AA2014–NiTi-based metallic composites through destructive and nondestructive evaluation. Russ J Nondestr Test 56(12):1064–1082
37. Zhang S, Kohlbrecher J et al (2013) Defect-induced Au precipitation in Fe–Au and Fe–Au–B–N alloys studied by in situ small-angle neutron scattering. Acta Mater 61(18):7009–7019

38. Zheng XG, Shi YN, Lu K (2013) Electro-healing cracks in nickel. Mater Sci Eng, A 561:52–59
39. Hsain Z, Pikul ZH (2019) Low-energy room-temperature healing of cellular metals. Adv Funct Mater 29(43)
40. Song H, Wang ZJ et al (2017) Self-healing of damage inside metals triggered by electropulsing stimuli. Sci Rep 7(1):7097
41. Kumar A, Paul SK (2020) Healing of fatigue crack in steel with the application of pulsed electric current. Materialia 14
42. Song H, Wang Z-J (2008) Microcrack healing and local recrystallization in pre-deformed sheet by high density electropulsing. Mater Sci Eng A 490(1–2):1–6
43. Wang F, Qian D et al (2019) Voids healing and carbide refinement of cold rolled M50 bearing steel by electropulsing treatment. Sci Rep 9(1):11315
44. Yang CL, Yang HJ et al (2018) Recovery of tensile properties of twinning-induced plasticity steel via electropulsing induced void healing. Scripta Mater 147:88–92
45. Tang Y, Hosoi A et al (2013) Effect of high-density electric current on the microstructure and fatigue crack initiation of stainless steel. Mater Trans 54(11):2085–2092
46. Zhou Y, Guo J et al (2004) Crack healing in a steel by using electropulsing technique. Mater Lett 58(11):1732–1736
47. Yu T, Deng D et al (2016) Crack healing in SUS304 stainless steel by electropulsing treatment. J Clean Prod 113:989–994
48. Ren X, Wang Z et al (2020) The plastic flow model in the healing process of internal microcracks in pre-deformed TC4 sheet by pulse current. Mater Des 188
49. Xu W, Yang C et al (2018) Microcrack healing in non-ferrous metal tubes through eddy current pulse treatment. Sci Rep 8(1):6016
50. Baumans XDA, Lombardo J et al (2017) Healing effect of controlled anti-electromigration on conventional and high-Tc superconducting nanowires. Small 13(26)
51. Putz B, Glushko O, Cordill MJ (2016) Electromigration in gold films on flexible polyimide substrates as a self-healing mechanism. Mater Res Lett 4(1):43–47
52. Tao J, Liew BK et al (1998) Electromigration under time-varying current stress. Microelectron Reliab 38(32):295–308
53. Danzi S, Schnabel V et al (2019) Rapid on-chip healing of metal thin films. Adv Mater Technol 4(3)
54. Balogh Z, Schmitz G (2014) Diffusion in metals and alloys. In: Laughlin DE, Hono K (eds) Physical metallurgy, 5th edn. Elsevier, Oxford, pp 387–559
55. Gale W, Totemeier T (1976) Diffusion in metals. In Smithells CJ (ed) Metals reference book, 5th edn.Butterworth-Heinemann, pp 860–939
56. Neumann G, Tuijn C (2008) Diffusion in Group VII Metals. In Neumann G, Tuijn C (Ed) Pergamon materials series. Pergamon, p 258
57. Neumann G, Tuijn C (2008) Self-diffusion and impurity diffusion in Group III metals. In Neumann G, Tuijn C (Ed) Pergamon materials series. Pergamon, pp 121–148
58. Lumley R (2007) Self healing in aluminium alloys. In van der Zwaag S (Ed) Self healing materials: an alternative approach to 20 centuries of materials science. Dordrecht, Springer Netherlands, pp 219–254
59. Wang H, Huang P, Li Z (2007) Crack and void healing in metals. In van der Zwaag S (Ed) Self healing materials: an alternative approach to 20 centuries of materials science. Springer Netherlands: Dordrecht, pp 255–277
60. Zhang HL, Sun J (2004) Diffusive healing of intergranular fatigue microcracks in iron during annealing. Mater Sci Eng, A 382(1):171–180
61. Gao KW, Qiao LJ, Chu WY (2001) In situ TEM observation of crack healing in α-Fe. Scripta Mater 44(7):1055–1059
62. Wei D, Han J et al (2006) A study on crack healing in 1045 steel. J Mater Process Technol 177(1):233–237
63. Chen C, Nagao S et al (2016) Self-healing of cracks in Ag joining layer for die-attachment in power devices. Appl Phys Lett 109(9):093503

64. Kovalenko O, Brandl C et al (2017) Self-Healing and shape memory effects in gold microparticles through the defects-mediated diffusion. Adv Sci (Weinh) 4(8):1700159
65. Nagai Y, Murayama M et al (2001) Role of vacancy–solute complex in the initial rapid age hardening in an Al–Cu–Mg alloy. Acta Mater 49(5):913–920
66. Djugum R, Lumley RN et al (2009) Enhanced fatigue resistance in a commercial Al-Cu-Mg alloy through underageing. In Proceedings 2nd international conference on self healing materials
67. Garrett GG, Knott JF (1975) Crystallographic fatigue crack growth in aluminium alloys. Acta Metall 23(7):841–848
68. Mahdavi Shahri M, Alderliesten RC et al (2014) Postponing crack nucleation in 2024 aluminium alloy by dynamic precipitation from the supersaturated state. Adv Mater Res 891–892:1577–1584
69. Mahdavi Shahri M, Ripoll MLR et al (2015) Exploring the option of fatigue life improvement of aluminium AA2024 via dynamic precipitation in the under-aged state. In van der Zwaag S, Brinkman E (eds) Self healing materials: pioneering research in the Netherlands. IOS Press.
70. Zhang Q, Zhu Y et al (2020) Training high-strength aluminum alloys to withstand fatigue. Nat Commun 11(1):5198
71. Barter SA, Molent L, Wanhill RJH (2012) Typical fatigue-initiating discontinuities in metallic aircraft structures. Int J Fatigue 41:11–22
72. Wanhill RJH (2007) Fatigue crack initiation in aerospace aluminium alloys, components and structures. In Proceedings of the first international conference on self healing materials, Noordwijk aan Zee, The Netherlands
73. Hannard F, Castin S et al (2017) Ductilization of aluminium alloy 6056 by friction stir processing. Acta Mater 130:121–136
74. Manuel, M.V., *Principles of Self-Healing in Metals and Alloys: An Introduction*, in *Self-Healing Materials*. 2008. p. 251–265.
75. He SM, van Dijk NH et al (2010) In situ determination of aging precipitation in deformed Fe-Cu and Fe-Cu-B-N alloys by time-resolved small-angle neutron scattering. Phys Rev B 82(17)
76. He SM, van Dijk NH et al (2010) Thermally activated precipitation at deformation-induced defects in Fe-Cu and Fe-Cu-B-N alloys studied by positron annihilation spectroscopy. Phys Rev B 81(9):094103
77. Laha K, Kyono J, Shinya N (2007) An advanced creep cavitation resistance Cu-containing 18Cr–12Ni–Nb austenitic stainless steel. Scripta Mater 56(10):915–918
78. Laha K, Kyono J, Shinya N (2011) Copper, Boron, and Cerium additions in type 347 austenitic steel to improve creep rupture strength. Metall Mater Trans A 43(4):1187–1197
79. Shinya N, Kyono J, Laha K (2006) Self-healing effect of boron nitride precipitation on creep cavitation in austenitic stainless steel. J Intell Mater Syst Struct 17(12):1127–1133
80. Zhang S, Langelaan G et al (2014) Preferential Au precipitation at deformation-induced defects in Fe–Au and Fe–Au–B–N alloys. J Alloy Compd 584:425–429
81. Zhang S, Kwakernaak C et al (2015) Self healing of creep damage by gold precipitation in iron alloys. Adv Eng Mater 17(5):598–603
82. Zhang S, Kwakernaak C et al (2015) Autonomous repair mechanism of creep damage in Fe-Au and Fe-Au-B-N Alloys. Metall Mater Trans A 46(12):5656–5670
83. Fang H, Versteylen CD et al (2016) Autonomous filling of creep cavities in Fe-Au alloys studied by synchrotron X-ray nano-tomography. Acta Mater 121:352–364
84. Versteylen CD, van Dijk NH, Sluiter MHF (2017) First-principles analysis of solute diffusion in dilute bcc Fe-X alloys. Phys Rev B 96(9):094105
85. Zhang S, Fang H et al (2016) Autonomous filling of grain-boundary cavities during creep loading in Fe-Mo alloys. Metall Mater Trans A 47(10):4831–4844
86. Fang H, Szymanski N et al (2019) Self healing of creep damage in iron-based alloys by supersaturated tungsten. Acta Mater 166:531–542
87. Versteylen CD, Sluiter MHF, van Dijk NH (2018) Modelling the formation and self-healing of creep damage in iron-based alloys. J Mater Sci 53(20):14758–14773

88. Fu Y, Kwakernaak C et al (2021) Surface precipitation of supersaturated solutes in a ternary Fe–Au–W alloy and its binary counterparts. J Mater Sci 56(8):5173–5189
89. Fu Y, Kwakernaak C et al (2020) Competitive healing of creep-induced damage in a ternary Fe-3Au-4W alloy. Metall Mater Trans A 51(9):4442–4455
90. Kazanjian SM, Wang N, Starke EA (1997) Creep behavior and microstructural stability of Al-Cu-Mg-Ag and Al-Cu-Li-Mg-Ag alloys. Mater Sci Eng, A 234–236:571–574
91. Kloc L, Cerri E et al (1996) Significance of continuous precipitation during creep of a powder mettallurgy aluminum alloy. Mater Sci Eng, A 216(1):161–168
92. Lumley RN, Morton AJ, Polmear IJ (2002) Enhanced creep performance in an Al–Cu–Mg–Ag alloy through underageing. Acta Mater 50(14):3597–3608
93. Li X, Liu W et al (2016) Radiation resistance of nano-crystalline iron: coupling of the fundamental segregation process and the annihilation of interstitials and vacancies near the grain boundaries. Acta Mater 109:115–127
94. Bai X-M, Voter AF et al (2010) Efficient annealing of radiation damage near grain boundaries via interstitial emission. Science 327(5973):1631–1634
95. Borovikov V, Tang X-Z, et al (2013) Coupled motion of grain boundaries in bcc tungsten as a possible radiation-damage healing mechanism under fusion reactor conditions. Nuclear Fusion 53:063001
96. Xiong F, Li M-F et al (2020) Microstructural evolution in amorphous-nanocrystalline ZrCu alloy under neutron irradiation. Acta Mater 182:18–28
97. Yang L, Li HY et al (2017) Structural responses of metallic glasses under neutron irradiation. Sci Rep 7(1):16739
98. Zhang S, Cizek J, et al (2020) Self healing of radiation-induced damage in Fe–Au and Fe–Cu alloys: combining positron annihilation spectroscopy with TEM and ab initio calculations. J. Alloys Comp. 817:152765
99. Zhang S, Yao Z, et al (2020) Irradiation damage and mechanical properties in Fe-Au and Fe-Cu model alloys under helium ion irradiation. Appl Surface Sci 504:144383
100. Wei D, Han J et al (2004) Simulation of crack healing in BCC Fe. Scripta Mater 51(6):583–587
101. Zhou G, Gao K et al (2000) Atomistic simulation of microcrack healing in aluminium. Modelling Simul Mater Sci Eng 8:603
102. Li S, Gao KW et al (2001) Molecular dynamics simulation of microcrack healing in copper. Comput Mater Sci 20(2):143–150
103. Wang MF, Du GJ, Xia DY (2013) Molecular dynamics simulation of microcrack healing in copper nano-plate. Key Eng Mater 531–532:454–457
104. Karpov EG, Grankin MV et al (2012) Characterization of precipitative self-healing materials by mechanokinetic modeling approach. J Mech Phys Solids 60(2):250–260
105. Versteylen CD, Szymański NK et al (2018) Finite element modelling of creep cavity filling by solute diffusion. Phil Mag 98(10):864–877
106. Xu GQ, Demkowicz MJ (2013) Healing of nanocracks by disclinations. Phys Rev Lett 111(14):145501
107. Meraj M, Pal S (2017) Healing mechanism of nanocrack in nanocrystalline metals during creep process. Appl Phys A 123(2):138
108. Li J, Fang QH et al (2015) Mechanism of crack healing at room temperature revealed by atomistic simulations. Acta Mater 95:291–301
109. Michalcová A, Marek I, et al (2018) Phase transformation induced self-healing behavior of Al-Ag alloy. Materials 11(2)
110. Kloc L, Spigarelli S et al (1997) Creep behavior of an aluminum 2024 alloy produced by powder metallurgy. Acta Mater 45(2):529–540
111. Sun WW, Fang H et al (2017) Linking surface precipitation in Fe-Au alloys to its self-healing potential during creep loading. Metall Mater Trans A 48(5):2109–2114

Advances in Self-healing Bituminous Materials: From Concept to Large-Scale Application

Jose Norambuena-Contreras, Quantao Liu, Alvaro Gonzalez, Alvaro Guarin, Nilo Ruiz-Riancho, Alvaro Garcia-Hernandez, Bastian Wacker, and Jose L. Concha

Abstract This chapter provides a comprehensive summary of the significant advances in self-healing bituminous materials from the engineering technologies associated with promoting the autonomous healing capability in the asphalt pavements to the large-scale applications.

1 Introduction

Bituminous materials are complex viscoelastic composites mainly used for road and airport pavement construction. Pavements are typically composed of layers that distribute the traffic loads evenly from the surface to the soil underneath. The top layer

J. Norambuena-Contreras (✉) · J. L. Concha
LabMAT, Department of Civil and Environmental Engineering, University of Bío-Bío, Concepción, Chile
e-mail: jnorambuena@ubiobio.cl

J. L. Concha
e-mail: jlconcha@ubiobio.cl

Q. Liu
State Key Laboratory of Silicate Materials for Architectures, Wuhan University of Technology, Wuhan, China
e-mail: liuqt@whut.edu.cn

A. Gonzalez
Department of Construction Engineering and Management, Pontificia Universidad Catolica de Chile, Santiago, Chile
e-mail: algonzav@ing.puc.cl

A. Guarin
Department of Civil and Architectural Engineering, KTH Royal Institute of Technology, Stockholm, Sweden
e-mail: alvaro.guarin@abe.kth.se

N. Ruiz-Riancho · A. Garcia-Hernandez
Nottingham Transportation Engineering Centre, University of Nottingham, Nottingham, UK
e-mail: Ignacio.RuizRiancho@nottingham.ac.uk

© Springer Nature Switzerland AG 2022
A. Kanellopoulos and J. Norambuena-Contreras (eds.), *Self-Healing Construction Materials*, Engineering Materials and Processes,
https://doi.org/10.1007/978-3-030-86880-2_4

is typically composed of asphalt mixture (95% w/w aggregates and 5% w/w bitumen), which is the most widely used road bituminous material. Aggregates give structural stability, while bitumen, a fluid with temperature-dependent viscosity, binds the aggregates together [1].

Once a road opens to traffic, each passing vehicle induces a small amount of damage to the pavement layers which, typically after 5–10 years, may cause damage (i.e. cracking). If asphalt pavement is too thin, cracks tend to initiate at the bottom and propagate to the surface. If pavement has an appropriate thickness, the cumulative effects of traffic, oxygen, UV radiation, thermal cycling and ambient moisture result in degradation of the bitumen and micro-cracks of the asphalt mixture, which causes cracks to appear at the surface in areas of high localized tensile stress [2]. Nevertheless, the cracking and degradation processes can be naturally mitigated by the natural drainage of bitumen into the micro-cracks, which effectively "self-heals" the asphalt layer [3]. Bitumen is highly viscous at the ambient temperature, so this self-healing is slow (days), and owing to the recurrent flow of traffic, the rate of crack growth may be faster than the healing, and hence macro-cracks often form, which cannot be "self-healed" due to their immense size.

So and considering the thermoplastic "autonomic healing" nature of bitumen, currently, two approaches have been used to promote self-healing capability in bituminous materials (including asphalt mixtures, mastics and bituminous binders): first, an *external self-healing* approach to reduce the viscosity of bitumen by increasing its temperature through by externally triggered heating using magnetic field technologies, such as electromagnetic induction heating (order of kilohertz) and microwave radiation (order of gigahertz) and, second, an *internal self-healing* approach that reduces the viscosity of bitumen by the effect of encapsulated rejuvenating agents added to the asphalt mixture during its manufacture process and activated under on-site load traffic conditions. See self-healing processes in Fig. 1.

Thus, to extend the road's service life and contribute to the development of more sustainable asphalt pavements, novel self-healing bituminous materials have been developed over the past years, turning this topic into an emerging field of study. As evidence, Fig. 1 shows the significant increase in the number of scientific papers published in the last years (period Jan. 2008–Dec. 2018) related to the hot topic "asphalt self-healing" according to the search results in Scopus database.

Furthermore, several large-scale projects (e.g. HEALROAD project) have been recently carried out as a consortium between Universities, Research centres and companies from Europe with the aim of underpinning the industrialization of on-site asphalt self-healing by induction heating technology. This chapter provides a

A. Garcia-Hernandez
e-mail: alvaro.garcia@nottingham.ac.uk

B. Wacker
BASt, Federal Highway Research Institute, Bergisch Gladbach, Germany
e-mail: wacker@bast.de

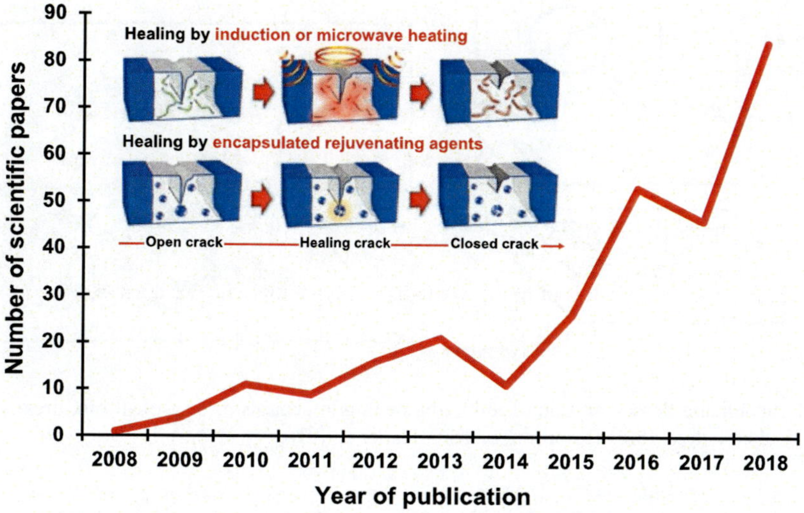

Fig. 1 Number of scientific papers published in the last years (period 2008–2018) searching in the Scopus database the keyword "*self-healing asphalt*". A total of 282 papers were considered

comprehensive summary of the significant advances in self-healing bituminous materials from the engineering technologies associated with promoting the autonomous healing capability in the asphalt pavements to the large-scale applications.

2 Induction Healing of Bituminous Materials with Fibres

Researches have demonstrated that temperature is the dominant factor influencing the self-healing properties of bituminous materials: an increase in the test temperature not only increases the self-healing rate but also shortens the total time needed for full healing [4, 5]. To enhance the self-healing capacity of asphalt mixture through increasing the temperature, an induction heating approach was developed in 2008 at the Delft University of Technology in the Netherlands. To do this, steel fibres were added into asphalt mixtures, and external induction heating was applied on the material to increase the healing capacity of asphalt mixture when micro-cracks occurred in the asphalt mastic [6–8].

During the induction heating process (see Fig. 2), asphalt mortar containing conductive fibres are exposed to a high-frequency alternating electromagnetic field, which can induce eddy currents in materials that are electrically and magnetically susceptible. The metallic fibres were heated by the induced eddy currents, and the heat energy diffuses into the bitumen to increase the temperature [9]. So, asphalt mixture can be healed quickly because bitumen behaves as a Newtonian fluid and flows very fast into the cracks under capillary force when its temperature is above its softening point [3, 10, 11].

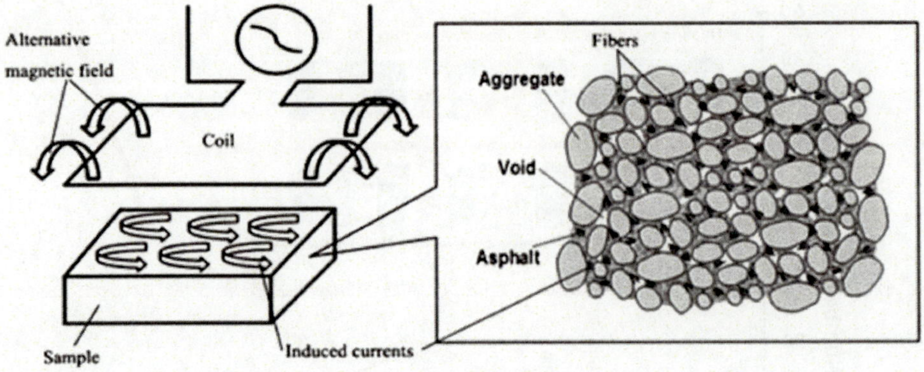

Fig. 2 Schematic demonstration of induction healing mechanism for asphalt mixtures containing metallic conductive fibres (Reproduced from Dai et al. [9])

2.1 Preparation of Asphalt Mixture Containing Steel Fibres

To make asphalt mixture electrically conductive and suitable for induction heating, conductive metallic fibres are usually incorporated into the asphalt mixture matrix. Commonly, steel wool fibres were directly added into the mixture during mixing without changing the gradation of the mixture [6–8]. Figure 3 shows the typical distribution of steel fibres into the asphalt mixture, where steel fibres form conductive paths and serve as the heating units. Long steel fibres are more useful to enhance the

Fig. 3 CT-scan reconstruction of the 3D spatial distribution of steel wool fibres into the porous asphalt mixture sample (Reproduced from Garcia et al. [11])

conductivity and induction heating speed of asphalt mixture [11]. However, they are complicated to mix and tend to form clusters, which will absorb too much bitumen and decrease the mixture's mechanical properties.

Short and thick steel wool fibres with diameter 70–130 μm and length 4.2 mm are recommended in the latest research [12]. This steel wool fibres are relatively easy to mix with the standard mixing procedure, and their optimal content is 6% by volume of bitumen. This steel wool fibre can increase Marshall stability, residual Marshall stability ratio, water stability, ravelling resistance, fatigue resistance and low-temperature properties of mixtures. Additionally, Liu et al. [13] in 2017 proved that a composite of steel fibres and steel slag (i.e. replacing a portion of mineral aggregates with steel slag) could enhance the induction heating speed, the heating homogeneity, and thus enhance the healing induction ratio of asphalt mixture.

2.2 *Heating Properties of Asphalt Mixture by Induction Heating*

In Li [12] asphalt mixture beams with a length of 200 mm, a width of 40 mm and height of 40 mm (steel wool fibres content was 6% by volume of bitumen) were heated by induction machine with an output power of 8.3 kW and frequency of 123 kHz. The settings and specifications of the induction machine determine the capabilities and the speed for heating the asphalt surface. In this section, possible implications for a certain machine are discussed. An infrared camera was used to monitor the temperature profile of the beam during heating. The distance between the coil and the surface of the beam was kept at 10 mm to obtain a high heating speed, according to the preliminary study published by Liu et al. [14].

From Li's [12] work, it was found that magnetic field intensity is not homogenous under the coil of the induction heating generator, which results in inhomogeneous heating in asphalt mixture (see Fig. 4). Even at the top surface of the asphalt sample, there is also a temperature gradient. Figure 4a shows the temperature distribution at the surface of the sample after heating 70 s, where most parts (80.7%) of the top surface of the sample have a temperature between 100 and 120 °C and the mean temperature is 107.2 °C (see Fig. 4b).

Otherwise, the vertical temperature distribution of the asphalt beam after 60 s induction heating was shown in Fig. 5. These results were published by Liu et al. [15]. In Fig. 5a, it can be seen from the infrared image that the temperature distribution showed an uneven heating effect in the vertical direction. The average temperatures of the top half and the lower part of the beam were 72.1 °C and 39.4 °C, respectively. It is worth noting that the ratio above 46.6 °C (around softening point of asphalt) was 55% (see Fig. 5b), meaning that half of the cracks can be healed at the optimal temperature.

The temperatures at different depths of the asphalt sample with different induction heating time, from 20 to 100 s, were shown in Fig. 6 [15]. It can be seen in Fig. 6

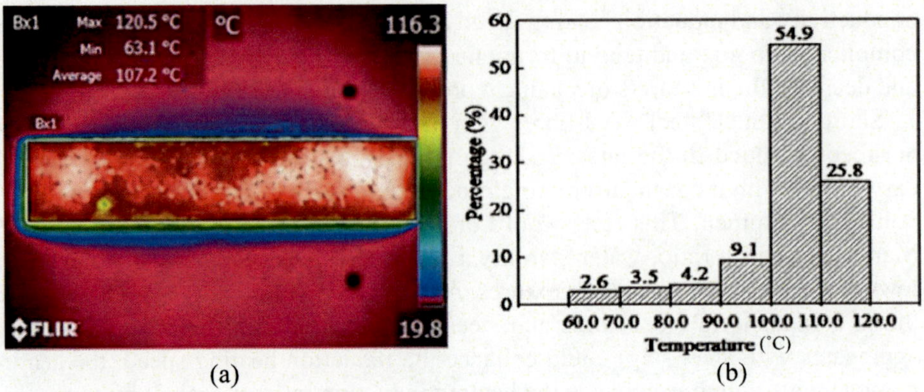

Fig. 4 Results of induction heating on the asphalt beam after heating 70 s: **a** Infrared image of the beam; **b** temperature distribution in the top-surface of the beam (Modified from Li [12])

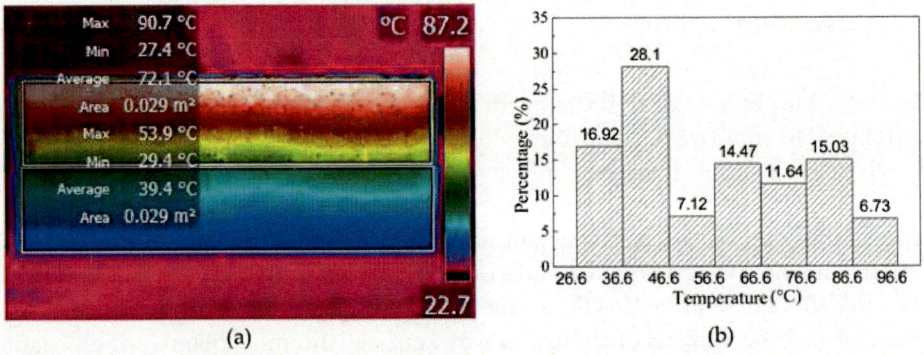

Fig. 5 Results of induction heating on the asphalt beam after 60 s: **a** Infrared image of the beam; **b** temperature distribution in the top-surface of the beam (Reproduced from Liu et al. [15])

that the temperature difference between the surface and the bottom of the samples increased with increasing heating time. The surface temperature reached 55.5 °C after 20 s heating, while the bottom temperature was only 27.5 °C, resulting in a temperature difference of 28 °C. The temperature difference between the surface and the bottom of the sample reached 92.3 °C after 100 s heating. It means that temperature distribution within the sample is much uneven. Also, it can be seen from Fig. 6 that the vertical temperature change rate increased from 0.35 °C/mm to 1.15 °C/mm with heating time increasing from 20 to 100 s. The considerable temperature difference will have a significant influence on the crack-healing behaviour of samples.

It can also be seen from Fig. 6 that the heating speed decreased gradually with the deepening of the depth. The temperature at a depth of 80 mm only increased by 14.9 °C after 100 s, which is not enough to activate the asphalt binder's healing. So, a sufficient healing depth exists after each heating time, and it is less than the thickness of the samples.

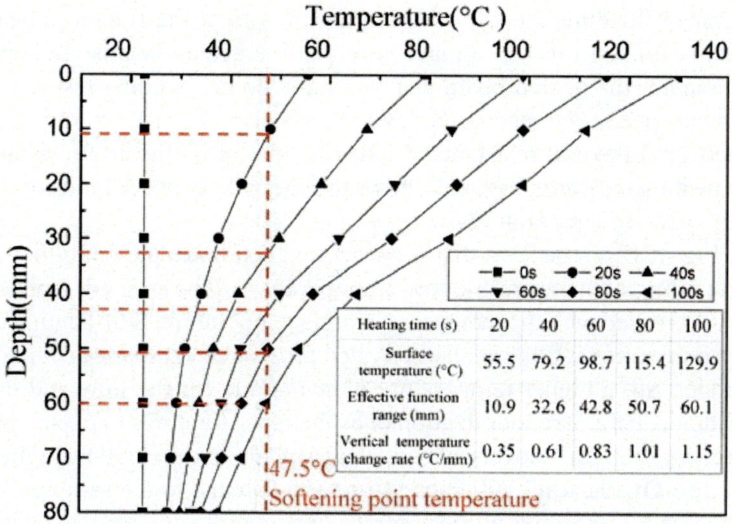

Fig. 6 Vertical temperature distribution of the asphalt samples containing steel wool fibres after different induction heating time (Reproduced from Liu et al. [15])

Otherwise, the healing efficiency of asphalt mixture also depends on the capillary flow speed of bitumen. The softening point of bitumen can be considered as an excellent healing temperature. The softening temperature of 70# base bitumen used in this research was 47.5 °C. Hence, the depth at which the temperature reached 47.5 °C was considered as the adequate healing depth of the sample and the sufficient healing depths at the different heating time were shown in Fig. 6.

From Fig. 6, it can be seen that the effective healing depth increased with the increase of the heating time. The effective healing depth reached 10.9 mm after 20 s heating, while increased to 60.1 mm after 100 s. However, the surface temperature of the sample reached 129.9 °C after 100 s heating, and the sample suffered serious deformation. As shown in previous research [13], the surface heating temperature should not be higher than 100 °C to avoid the swelling of the binder and the excess expansion of the mixture; it can be assumed that the maximum effective depth of asphalt mixture with induction heating is around 42.8 mm with 60 s heating. Extending the heating time further will result in excess expansion deformation of the mixture.

2.3 *Strength Recovery of Asphalt Mixture by Induction Heating*

In Liu et al. [14], the test procedure for measuring induction heating induced healing of fracture damage in asphalt beams (250 mm × 35 mm × 30 mm) consisted of three steps: (1) Fracture the asphalt beam in three points bending test at −10 °C with a displacement loading speed of 50 mm/min; (2) Heat the fractured beam and let the heated beam rest for 4 h to make sure that the heated beams are cooled down; and (3) Fracture the healed beam again in three points bending test with the same displacement speed. The recovered flexural resistance of the fractured beam defined as the recovered flexural resistance divided by the initial flexural resistance can be used as a healing indicator. Figure 7 shows the strength recovery ratios of the samples at different heating temperatures.

From Fig. 7, it can be seen that the fractured beam sample could not recover its strength at room temperature and lower temperatures. The strength recovery ratio of the beams increased with the increasing heating temperature. Self-healing of bitumen and asphalt mixtures is a thermally activated process of capillary flow and diffusion of the binder. So, a higher temperature is favourable for the flow and diffusion of bitumen molecules and randomization of asphaltene structures to regain the strength. Therefore, the strength recovery of a cracked asphalt mixture will be higher at higher temperatures. Overheating will cause structural damage to the sample. Swelling of the mortar can be observed in the samples heated to 95 °C. This explains why the healing rate at 95 °C is lower than at 85 °C. Binder drainage may also occur in these overheated samples and that can also offset the beneficial effect of heating. Based on this result, it is concluded that the optimal heating temperature is 85 °C for the sample to obtain the highest fracture resistance recovery. The maximum strength ratio of the fractured beam is 77.9%. It means that the induction healing is not complete (i.e. 100%). The reason for the incomplete healing mainly lies in the fact that some aggregates were broken during the test, which cannot be healed. Another possible reason for the limited healing could be the temperature gradient through the depth of the sample. The sample is fully damaged through its height, but induction heating tends only to heal the damage in the top part of the beam, where the temperature is higher than in the lower part after induction heating.

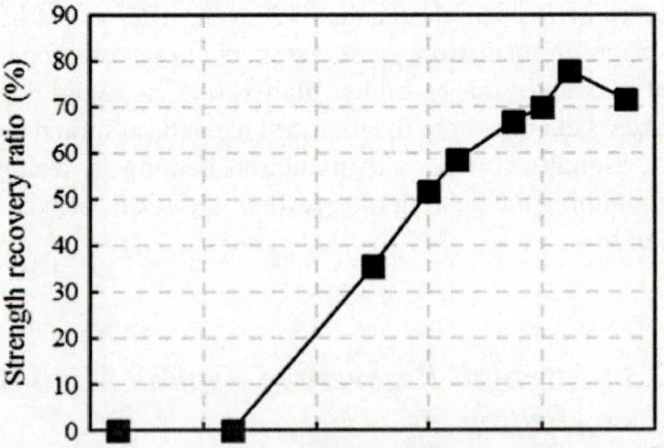

Fig. 7 Strength recovery ratios of the asphalt mixture samples at different heating temperatures (Modified from Liu et al. [14])

2.4 Fatigue Life Extension of Asphalt Mixture by Induction Healing

In Liu et al. [14], the experimental method of testing the induction heating activated healing of fatigue damage in asphalt beams consisted of three steps: (1) fatigue testing at 15 °C, 10 Hz and with 600 micro-strain was performed on asphalt beams (380 mm × 63.5 mm × 50 mm) until the complex modulus dropped to half of its initial value; (2) the fatigue damaged beams were heated to different temperatures; and (3) finally the heated beams were cooled to 15 °C and tested again under fatigue until the complex modulus dropped to the same value as the first fatigue testing. The second fatigue life is a healing indication caused by resting and induction heating. The healing index fatigue life extension ratio can be defined as the second fatigue life divided by the original fatigue life.

To quantify the effect of heating temperature on the healing rates of asphalt mixture, the fatigue life extensions of the fatigue damaged samples with different heating temperature were investigated. Figure 8 shows the induction healing ratios (fatigue life recovery ratios) of the samples at different heating temperatures [14].

As shown in Fig. 8, the healing ratios of the samples are highly temperature-dependent. The healing ratio increases dramatically with the increasing heating temperature due to the faster flow of bitumen and obtains the maximum healing ratio before decreasing. The reason for this decrease can be attributed to the geometry damage of the sample caused by overheating. In this case, the asphalt mortar suffered excess expansion due to overheating, and the swelling problem showed up in the sample. The optimal induction heating temperature for the beams 87 °C and the maximum fatigue life ratio of the samples is 56.2%.

Likewise, in Liu et al. [16], the possibility of multiple instances of induction heating also was examined to show that induction heating technology could be repeated when cracks return after healing. To prove it, a strain amplitude of 300 microstrains at a frequency of 8 Hz was applied to the porous asphalt mixture beams for 50,000 cycles. Then, asphalt samples were induction heated to 85 °C and rested for

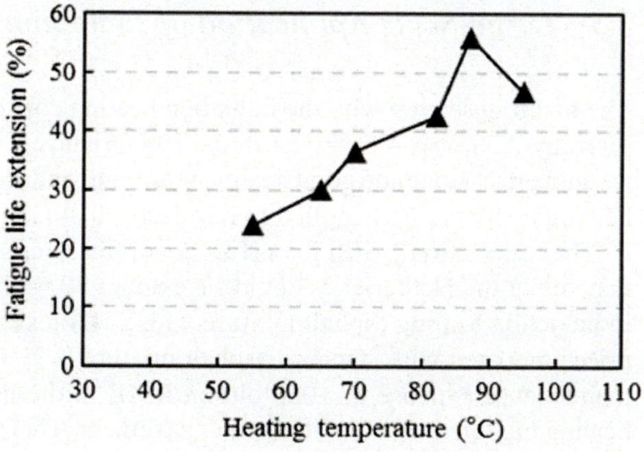

Fig. 8 Fatigue life extension ratios of the asphalt mixture samples versus heating temperature (Modified from Liu et al. [14])

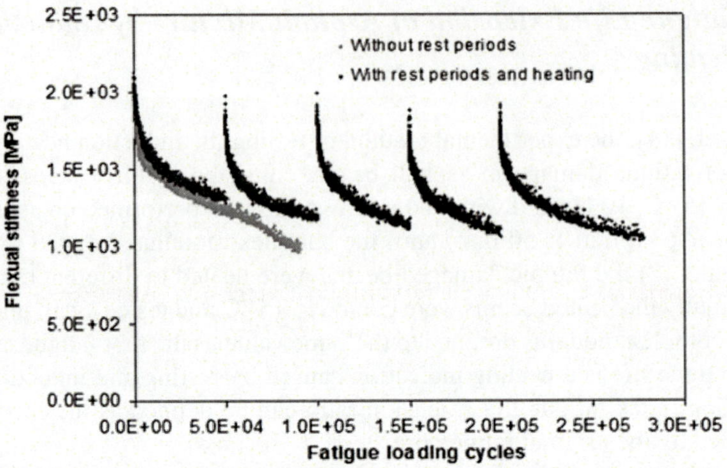

Fig. 9 Fatigue life extension of the porous asphalt sample caused by multiple times induction heating (Reproduced from Liu et al. [16])

18 h for the first time. After that, another 50,000 cycles of fatigue loading were applied to the beams, followed by a second heating-and-resting process. The damaging, heating-and-resting and re-damaging process was repeated four times. Finally, the beam fatigue life was measured. Figure 9 compares the original fatigue curve and the modified fatigue curve for the sample with multiple instances of induction heating to show the fatigue life extension caused by induction heating [16].

In Fig. 9, the original fatigue life of the sample is 95,700 cycles. With four instances of damage loading of 50,000 cycles followed by four instances of induction heating and resting, the modified fatigue life is 277,720 cycles, which is 2.9 times the original fatigue life; the fatigue life extension ratio was 190% in this case. Even though this research does not fully optimize when to heat the pavement and heating frequency, multiple heating instances can extend the fatigue life of the mixture.

2.5 Large-Scale Application of Induction Heating on Roads

The first trial section with the induction heating concept was constructed in 2010 on highway A58 near Vlissingen in the Netherlands. This trial section showed better ravelling resistance and good healing ratios with induction heating. The first induction heating treatment was applied to it in June 2014 (see Fig. 10).

The test section is still in perfect condition after serving eight years. It was estimated that the Netherlands could save about 90 million euro annually by investing in induction healing asphalt mixture with a 50% extended life span and twice of the price compared with standard asphalt mixture [17]. Furthermore, China will save a maintenance expense of 1000 billion RMB in the life span by applying induction healing mixture in 10% of its asphalt pavement. Therefore, induction healing asphalt

Fig. 10 Large-scale application of induction heating on Dutch Highway A58 in the Netherlands

mixture, being a smart material and an advanced maintenance concept for asphalt pavement, is promising to significantly improve the quality of the pavement, reduce the maintenance activities and extend its service life over time.

3 Induction Healing of Asphalt Modified with Nanoparticles

Asphalt healing is described as micro-cracks being refilled, to a certain degree, with bitumen seeping from the zone around the crack during summer due to high temperatures. This effect is often referred to as "self-healing" because it is usually induced by sun radiation, without any human action. It may make sense to add healing promoters as nanoparticles to the bitumen, to increase its healing [18].

Among other nanomaterials, carbon nanotubes, nanoclay particles nanoceria, coal tar, nano-TiO_2, nano-SiO_2, nano-calcium carbonate and rubber nano-powders have been utilized to improve tensile strength, Marshall stability, water stability, temperature performance, ageing properties, rheology, rutting and cracking resistance of asphalt mixtures [18]. Mostly, the focus has been on the application of nanoparticles to change rheology of bitumen, especially at high temperature, but relatively little work has been done regarding cracking and healing [19].

Microstructures and molecular components of nanomodified bitumen have been investigated with innovative technologies such as atomic force microscopy (AFM); scanning electron microscopy (SEM), X-ray diffraction and Fourier transform infrared spectroscopy [20]. Nanoparticles may improve the mechanical performance; enhance thermal conductivity and/or electrical conductivity of bituminous materials, boosting their healing potential.

A well-documented and crucial issue is that, due to their higher surface area per mass unit, nanoparticles for construction industry may cause severe health and environmental issues; for instance, regarding nanomaterials used for modification of bitumen, titanium dioxide TiO_2 nanoparticle inhalation has been linked to severe lung disease [21, 22]. Also, nanomaterials can have severe toxicity risks; carbon nanotubes (CNTs) may induce malignant mesothelioma, fatal cancer that affects lungs and chest cavity, which poses an asbestos-like hazard for human health [23].

3.1 Effect of Nano-Modification on Asphalt Healing

Qiu et al. [24] evaluated the effect of nanoparticles (full-vulcanized rubber powders called NanoA and NanoB, provided by Beijing Research Institute of Chemical Industry) on healing of bitumen. They performed ductility, dynamic shear rheometer DSR and direct tension tests to assess healing of bitumen; they found it a promising path to improve the self-healing capability of bituminous binders.

Organoclay was utilized as a nano-modifier for bitumen. Dynamic shear rheometer was used to measure dissipated energy, the ratio of dissipated energy change and plateau value energy of organoclay-modified binders during fatigue tests. The effect of rest periods during the time sweep tests on the fatigue life of neat and modified binders was also studied. Based on three healing indexes, it was concluded that organoclay-modified binders exhibited better healing than neat binders [25].

Fang et al. [26] analysed the preparation and processing of nanomodified asphalt binder (dispersion) and some effects of nanomaterials on bitumen, including compatibility, stability, ageing, among other factors. They also pointed out the lack of understanding about shearing speed, temperature and mixing time.

Jeoffroy et al. [27] introduced a novel method to accelerate the heating of bitumen by stimulating pre-embedded iron oxide nanoparticles using an external oscillating magnetic field (hyperthermia); the fast decrease of the bitumen viscosity, and so the closing of micro-cracks in the bitumen (see Fig. 11). More homogeneous dispersion of nanoparticles was accomplished by adsorbing oleic acid on their surface; thus, a uniform temperature increase throughout the material was induced.

Magnetic oxide nanoparticles are not prone to corrosion; furthermore, heating can be adjusted according to the size, shape and concentration of nanoparticles. They investigated three commercially available iron oxide nanoparticles with particle sizes ranging from 20 to 100 nm. They established that nanoparticles with an average crystallite size of 50 nm exhibit the highest magnetic hysteresis losses. Then the powder is the most efficient in quickly increasing the temperature of the bitumen specimens. Radial cracks created by Vickers indentation of bitumen samples at low temperature, healed in few seconds, using the proposed hyperthermia effect [27].

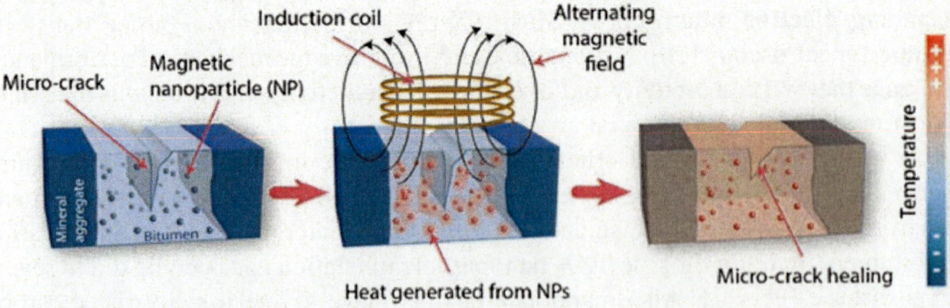

Fig. 11 Schematics illustrating the proposed crack-healing method for bituminous materials using magnetic nanoparticles embedded in bitumen (Reproduced from Jeoffroy et al. [27])

Multiwall carbon nanotubes CNT (average diameter 9.5 nm and length 1.5 μm) and montmorillonite nanoclay NC can enhance the healing properties of bituminous mastics, extending their lifespan. Healing potential was evaluated differentiating between recovery of fatigue loading induced damage and other artefacts that can influence the response of the material. Healing tests were carried on with dynamic shear rheometer (parallel plates); it should be noticed that both nanoparticle peculiarities and testing conditions have a substantial effect on the results [28].

3.2 Advances on Nano-Modificated Asphalt Healing

Wang et al. [29] utilized nano-inorganic/organic hybrid shells to manufacture self-healing microcapsules containing asphalt rejuvenator. Shells were composed of two layers: cross-linked methanol-modified melamine–formaldehyde (MMF) resin for the internal one while the external material was methanol-modified MMF resin and nanoparticles of calcium carbonate (nano-$CaCO_3$; 20 nm average particle size) (see Fig. 12). They found that the nanoparticles on microcapsules reduced the deformation of shells by nanoindentation tests; they also concluded that the addition of nano-$CaCO_3$ made the microcapsules more resistant to aggressive temperature changes. Microcapsules survived in the bitumen with a temperature of 200 °C.

Carbon nanotubes were proposed as novel additives for bituminous materials suitable for microwave radiation heating to promote self-healing. Bitumen samples modified with different contents of multi-walled nanotubes MWNTs were characterized with dynamic shear rheometer DSR and scanning electron microscope SEM, among other tests. Healing capability was assessed with tensile tests on bitumen and indirect tensile tests of asphalt mixture (Marshall specimens); the results showed that carbon nanotube modified bitumen and asphalt mixtures had better healing performance [30].

As well, Ganjei and Aflaky [31] investigated if the addition of nanosilica (average particle size of 60 nm) to modified binder with Styrene–Butadiene–Styrene SBS would foster the healing capability of asphalt mixtures using the Taguchi design of experiment (DOE) method. Based on Superpave indirect tensile tests, including repeated loading and rest periods, they demonstrated that the combination of

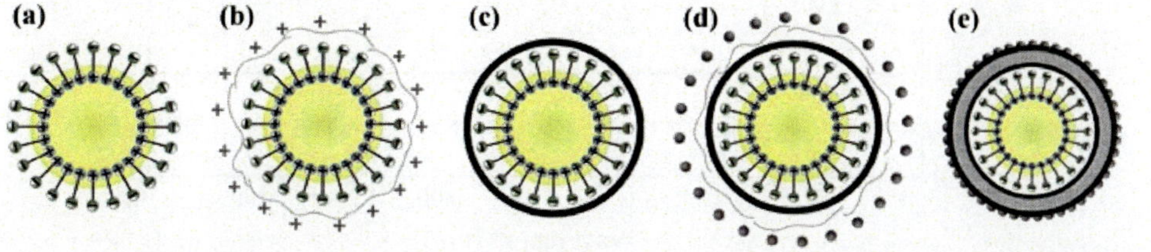

Fig. 12 Microcapsules fabrication process, (**a–c**) the inside-layer formation of shell, and (**d**, **e**) the out-layer shell with nano-$CaCO_3$ composite structure (Modified from Wang et al. [29])

nanosilica and SBS substantially enhanced self-healing of asphalt mixtures. The asphalt mixture self-healing mechanism was further explored with scanning electron microscope. An optimum condition for maximum healing potential was suggested.

Wang et al. [32] utilized flake graphite and exfoliated graphite nanoplatelets xGNP (particle diameter of 25 μm, and an average thickness of approximately 15 nm) to improve both light absorption and thermal conductivity of asphalt mixtures to accelerate light healing (see Fig. 13). The asphalt specimens were heated at the surface with light absorption; the added graphite particles enhanced the near-field infrared light absorption, then the heat was transferred to a deeper zone in the sample. Heat transfer initiates the crack healing process, a combination of capillary flow, the gravity of liquid asphalt filled in the crack and friction between the flowing asphalt and aggregate particles.

The displacement evolution in the fracture zone during the light healing process was assessed with digital image correlation DIC; those results suggested enhanced healing response of graphite-modified asphalt mixtures. Cyclic fracture-light healing tests were done on graphite-modified asphalt mixture beams (5% flake graphite and 2% xGNP); it was seen that the graphite-modified asphalt specimens have significantly better healing performance. The recovered strength was promoted by the diffusion of asphalt molecules and rebinding of asphalt and aggregate particles [32].

Jeoffroy et al. [33] further investigated the feasibility of using magnetically responsive iron-based particles as additives to induce heat on asphalt, and consequently, enhance healing potential. Cast steel particles were tested in eight different size ranges: 0.05–0.18, 0.13–0.43, 0.30–1.00, 0.71–1.40, 1.20–1.70, 3.00, 4.00 and 6.00 mm. The healing capability was verified by the recovery of samples during a double torsion test before and after exposure to the alternating magnetic field AMF.

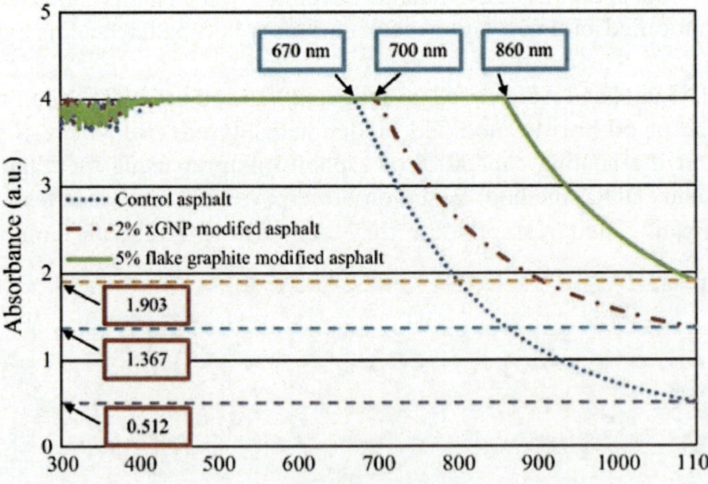

Fig. 13 Light absorption of control asphalt, flake graphite-modified asphalt and xGNP modified asphalt (Reproduced from Wang et al. [32])

They observed that the size and chemical composition of the iron-based particles play a fundamental role in the heating mechanism, kinetics and the maximum temperature increase on the surface of the asphalt. The authors also concluded that temperature on the asphalt increased with particle size until reaching a maximum at a critical size around 1–2 mm, which is the critical size corresponding to the particle skin depth. Heating of asphalt was less effective when metallic particles were larger than 1–2 mm since the induced eddy currents are limited to the skin close to the particle surface (Fig. 14). Design guidelines for the use of iron-based particles for crack healing of bituminous materials were provided.

Yoo et al. [34] evaluated the effect of carbon fibres (CFs), carbon nanotubes (CNTs, 5–20 nm particle diameter) and graphite nanofibers (GNFs, 50–200 nm particle diameter) on mechanical properties (Marshall stability, indirect tensile strength, dynamic stability and flexural performance) and self-healing behaviour of asphalt mixture. The healing level was calculated as the ratio of flexural strengths of self-healed and undamaged virgin test specimens. The induction heating (heat generator) due to the added carbon materials led to about 40% recovery of the original flexural strength for the specimens with 0.5% GNFs and CFsn.

Li et al. [35] found that two-nanometer microwave-absorbers (carbon nanotubes CNTs and graphene) not only expedite the heating process considerably under microwave radiation (Fig. 15), but also extend the time for diffusion and healing of modified bitumen in comparison with basis bitumen. This is important to enhance healing potential, since the self-healing process of bitumen takes some time and gradually diffuses across the entire crack interface. The authors measured the initial self-healing temperatures of basis and modified bitumen; they also concluded that CNTs and graphene-enhanced the high-temperature performance of original binder, but had detrimental effects on low-temperature cracking performance. The healing index (ratio of maximum breaking force of healed sample and maximum breaking force of original sample) of original bitumen, graphene-modified bitumen and CNTs

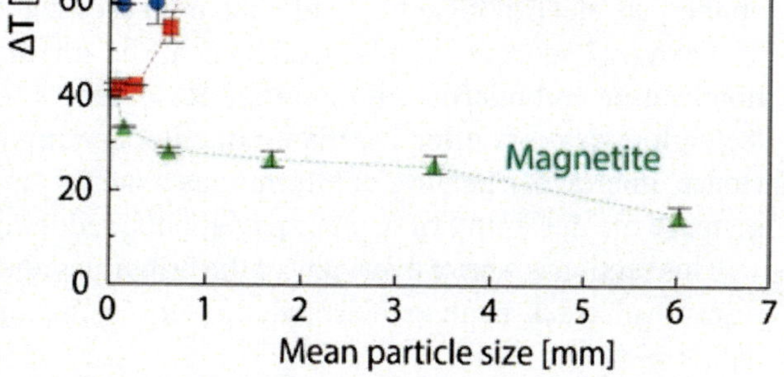

Fig. 14 Temperature changes recorded at the surface of asphalt samples containing 3 vol% of particles versus the mean particle size (Reproduced from Jeoffroy et al. [33])

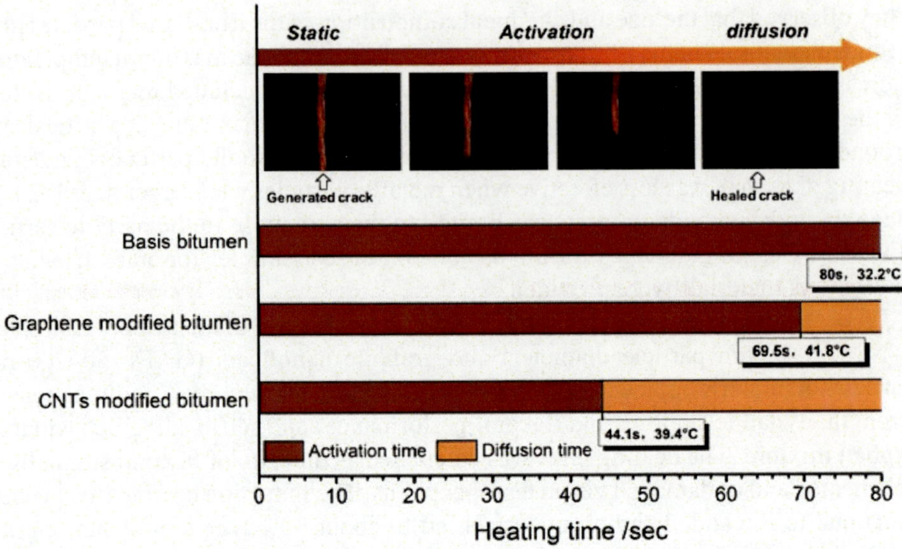

Fig. 15 Apparent healing effect and healing stage of different bitumen during heating process (Reproduced from Li et al. [35])

modified bitumen was 13.89%, 69.15% and 75.20%, respectively, after 12 h of rest in the cracking–healing–re-cracking test, indicating that modified bitumen exhibited better healing performance than basis bitumen.

4 Microwave Healing of Bituminous Materials

In the last decade (see Fig. 1), researchers developed the principle of self-healing asphalt by externally increasing bitumen temperature. When heated, these asphalt mixtures have the capability of sealing or healing open cracks [36, 37]. To achieve more efficient heating, this type of asphalt mixtures typically contain steel fibre or other additives that increase thermal absorption, electrical conductivity and dielectric properties [38, 39]. Hence, the heat transfers efficiently throughout the mixture, and the healing is more efficient than conventional mixtures without additives.

To artificially heat and heal this type of asphalt mixtures, an external source applies an electromagnetic field that increases the fibre or additive temperature. The external sources mainly used to heat the mixtures are electromagnetic induction heating and microwave radiation. Researchers [40] have found that microwave technology is more effective than induction heating to heal cracks in asphalt roads. Hence, microwave healing of bituminous materials is a promising technique that can increase the durability of asphalt pavements, reducing maintenance costs.

This section is a brief overview of the principles and main results obtained in recent research on crack-healing of asphalt mixtures using microwave radiation heating. The section describes the fundamentals of microwave heating of composite materials, the crack-healing mechanism of asphalt mixtures, the most common laboratory tests used to measure healing of asphalt mixtures, the additives normally added to increase the microwave heating capabilities of asphalt mixtures, and the effects of crack and healing with microwaves on the strength of asphalt mixtures.

4.1 Microwave Heating Principle

4.1.1 Polarization Mechanisms

Microwave heating is an industrial process developed during the 1940s. The pioneer in the field of dielectric materials and their applications was von Hippel [41], who studied the properties of many organic and inorganic materials under microwave radiation in the frequency between 10^2 and 10^{10} Hz [42]. Since then, many industrial processes commonly use electromagnetic fields for heating different types of materials. Radiofrequency heating applies waves below 100 MHz, using open wire circuits. The microwave frequencies are higher than radiofrequency, above 500 MHz; however, wired circuits do not work at this high-frequency range. Instead, the electromagnetic waves transfer power to the applicator containing the material. The principle of microwave heating lies in the ability of the electric field to polarize the charges in the heated material and the inability of this polarization to follow extremely rapid reversals of the electric field produced by microwaves. The lag between the applied electromagnetic field and the material's charge dissipates power, producing heat [42]. There are four components of polarization related to microwave heating:

- Electronic polarization, i.e. the displacement of electrons around the nuclei.
- Atomic polarization, i.e. the relative displacement of atomic nuclei because of the unequal distribution of charge in molecule formation.
- Interfacial polarization (Fig. 16a), arising from charge build-up in interfaces between components in heterogeneous systems.
- Orientation polarization (Fig. 16b), occurring in materials with permanent dipoles due to the asymmetric charge distribution in a molecule which tends to re-orientate under the influence of a changing electric field (e.g. water molecules).

The last two components of polarization (interfacial and reorientation) are the most important mechanisms of high-frequency heating. This is the technique used in most of the research conducted on crack-healing of asphalt mixtures.

In addition to the polarization effects described above caused by microwaves, the materials heat through conduction effects, e.g. the redistribution of charge particles forming conducting paths.

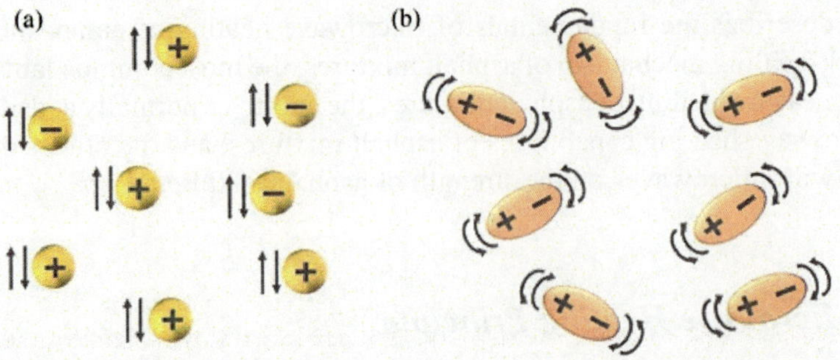

Fig. 16 Schematic representation of the two polarization mechanisms that are the basis of high-frequency heating: **a** interfacial, **b** reorientation (Inspired by Metaxas and Meredith [42])

4.1.2 Electrical Permittivity

Electrical permittivity is a fundamental property that describes how much the molecules of a material oppose an external electric field. The relative permittivity of a material compared to the permittivity of a vacuum (no electrons present) is known as the dielectric constant of the material [43]. In other words, it is an expression of the extent to which a material concentrates electric flux. However, permittivity says nothing about the power absorption of the material. The relative dielectric constant ε^* of a material is defined as:

$$\varepsilon^* = \varepsilon' - j\varepsilon'' \quad (1)$$

where the real part ε' is the permittivity of the material, and the imaginary part, ε'', is the dielectric loss factor. By electromagnetic theory, it is demonstrated that power dissipation p of a material heated by microwave with frequency f is given by:

$$p = 2\pi f \varepsilon_0 \varepsilon'' (E_i)^2 \quad (2)$$

where E_i is the magnetic field (V/m), and ε_0 is the permittivity of free-space [43].

Researchers have measured the dielectric properties of asphalt pavements using a roller mountable density sensor [44] and applying frequencies within the range from 100 Hz to 12 GHz. Results showed that permittivity and dielectric loss depend on the frequency and temperature of the pavement and that the higher the pavement density, the higher the permittivity. Results showed that permittivity slightly increases with temperature and moisture. Also, they found that moisture strongly increases permittivity and loss at low frequencies and only slightly at microwave frequencies. Researchers concluded that the penetration depth of microwaves in asphalt pavement is about 12–14 cm at 8 GHz and only 4 cm at 30 GHz, indicating that heating thick asphalt pavements are possible at low microwave frequencies. These findings are useful for the application of microwaves to asphalt in the field.

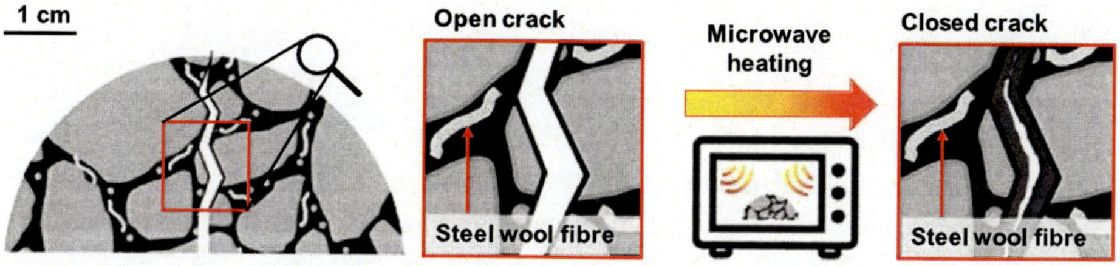

Fig. 17 Self-healing of asphalt mixtures containing steel wool fibre by microwave radiation (Modified from Norambuena-Contreras and Gonzalez-Torre [47])

4.2 Healing Mechanism

The principle of self-healing asphalt mixtures by bitumen temperature increase was used to create an asphalt mixture with crack-healing properties [3]. In these mixtures, metals, usually steel wool fibres (Fig. 17), are added because they absorb and conduct more thermal energy than bitumen and aggregates, improving the electrical conductivity of the asphalt mixtures [38]. To artificially heat and heal this type of fibre-reinforced asphalt mixture, an external electromagnetic field, such as those applied by electromagnetic induction or microwave radiation, is used to increase the fibre temperature. Later, the fibre heat transfers to the bitumen and aggregates, reducing the bitumen viscosity and therefore repairing open cracks [45, 46].

One disadvantage of adding steel wool fibres is that they tend to cluster in balls and increase the air void content of the mixtures. This increase in porosity is associated with a reduction of the mechanical properties [39]. Other researchers [48] have only evaluated the effect of adding metallic waste in the heating of asphalt mixtures, without assessing the crack-healing capabilities. Recently, researchers tested the healing mechanism of mixtures with different content of reclaimed asphalt pavement (RAP) with promising results [49], although increasing RAP contents decrease the healing capability of the mixtures. A number of waste materials are used in asphalt mixtures without healing purposes, with the aim of replacing virgin materials with waste, or improving other properties of the mixtures [50–54].

4.3 Measurement of Healing

The simple three-point bending test is typically conducted on semicircular samples to calculate the flexural strength of asphalt. This test has been previously used to assess the healing capabilities of asphalt mixtures [39, 40, 47, 54]. In these tests, the test specimen is placed onto two supporting rollers, and a third loading roller is positioned at the midpoint of the semicircular arch of the sample (see Fig. 18).

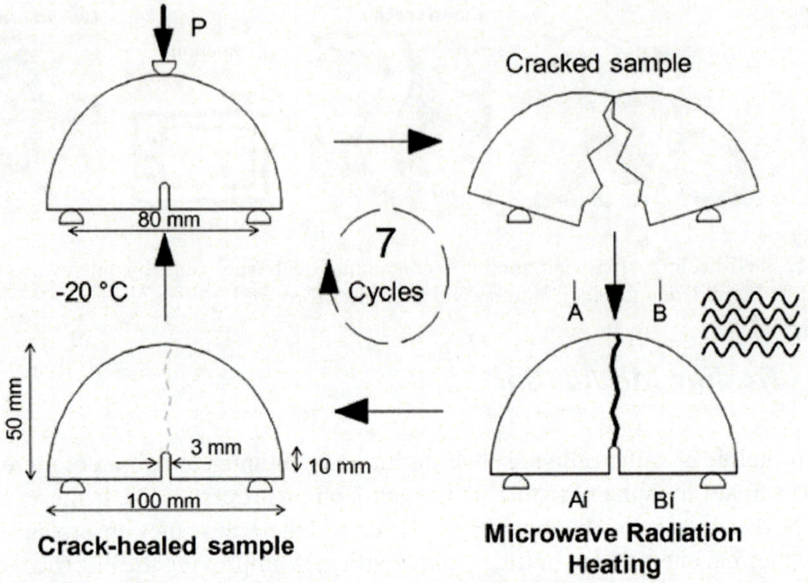

Fig. 18 A sequence of a crack-healing cycle: (step 1) loading a semicircular specimen (step 2) cracking (step 3) heating by the application of microwaves, and finally (step 4) healing the crack once the bitumen cools down (Reproduced from Gonzalez et al. [49])

The test starts by applying a monotonic load that induces tensile stress at the upper tip of the vertical notch located at the bottom of the sample. The notch guides the crack propagation in the upper direction during the test. Generally, the three-point bending tests are conducted at −20 °C to achieve brittle behaviour of the mixtures and a single crack throughout the sample. However, some authors report these tests at higher temperatures [55]. In addition to the three-point bending tests, some researchers have used the indirect tensile strength test to evaluate cracking resistance of asphalt mixtures healed by microwave heating [56].

The microwave healing of the asphalt mixtures is generally assessed with the healing ratio, i.e. the ratio between the three-point bending strength F_i of the sample after the ith healing cycle, and the initial strength F_0 of the same sample before healing [57]. The monotonic universal testing machine loads the samples up to failure, recording the peak load. Once the bending test finished, cracked asphalt samples are withdrawn from the loading machine and are left at room temperature for at least 3 h until they reach room temperature.

4.4 Metallic Additives for Microwave Heating in Asphalt Mixtures

4.4.1 Steel Wool Fibres and Steel Shavings

Steel wool fibres (Fig. 19a) are the most used additives reported in the literature for the fabrication of asphalt mixtures with self-healing purposes by means of external heating. Based on Norambuena-Contreras et al. [39, 40], these metallic fibres are composed of low-carbon steel with a density of approximately 7.180 g/cm^3.

These additives are added in small proportions to the mixtures, normally between 2 and 8% of the volume of the bitumen, where the diameter and length of the steel fibre range between authors. For example, Gonzalez et al. [58] used steel wool fibre with average diameter of 0.133 mm, and an initial length range of 2–14 mm. Figure 19a' shows a scanning electronic microscope (SEM) image of a steel wool fibre surface used for the fabrication of asphalt mixtures with crack healing capabilities. Although fibre is an industrial product, the SEM shows defects on the surface.

Steel shavings are an industrial waste usually formed by ferritic steel with densities higher than steel fibre (7.980 g/cm^3). The shape of steel shavings is different from steel fibre. For example, Norambuena-Contreras et al. [57, 58] reported an average width of 1.31 mm and initial length within the range 3–21 mm. Some shavings had helicoidal shape while others are long curled metal particles (Fig. 19b). SEM of metal shavings (Fig. 19b') shows defects on the edges and surfaces of the shavings, which

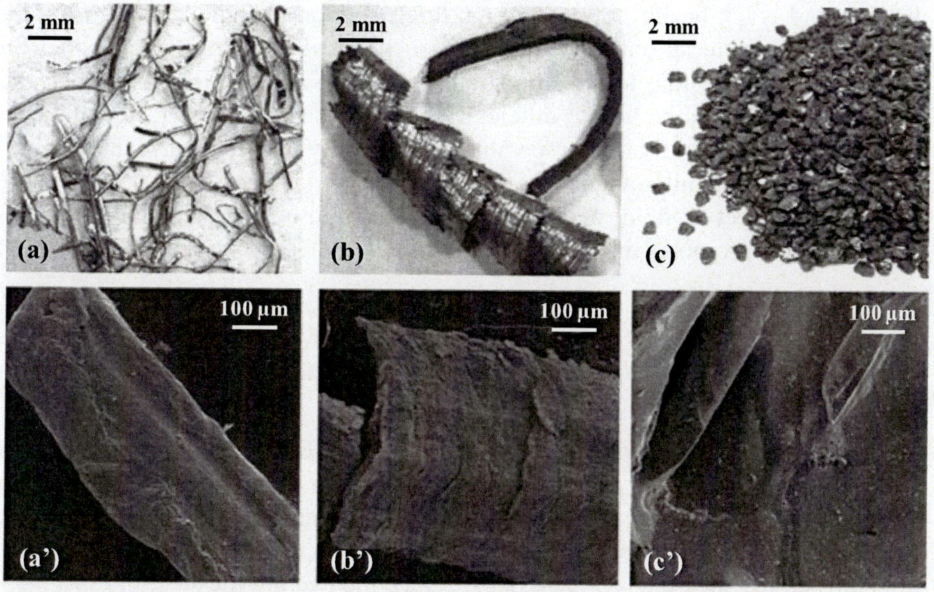

Fig. 19 Microscopy images of several metal additives used in asphalt mixtures for microwave healing: **a** steel fibre, **b** metal shavings, **c** silicon carbide

is expected since shavings are obtained from metal turnery, a process in which metal tools and saws apply high shear stresses on metals. Similarly to steel wool fibres, the shavings content by the total volume of the bitumen in asphalt mixtures with crack healing purposes ranges from 2 to 8% [39, 40, 49, 57, 58].

4.4.2 Metallic Powders and Aggregates

Researchers have used metallic powders to increase the microwave dielectric properties of asphalt mixtures. Zhu et al. [59] studied the effect of ferrite content on the mechanical properties and heating effect of asphalt mixtures. They found that the optimal microwave heating time and optimal healing opportunity of an asphalt mixture used in open grade friction course with 5% ferrite were 120 s and 50% damage degree, respectively. In addition, the researchers found that the longer the healing time, the better the healing effect, but the influence of the healing time is smaller, compared with the heating time. Sun et al. [60] used steel slag aggregate to promote asphalt heating by microwaves. The particle diameter of the steel slag ranged between 9.5 and 13 mm, the apparent specific gravity was 3.421, water absorption was 1.4% and the Los Angeles abrasion test was 14.5%.

4.4.3 Ceramic Additives with Good Dielectric Properties

Researchers have reported at least two types of carbon additives for the microwave heating and crack-healing of asphalt mixtures. Silicon Carbide (SiC) is a synthesized ceramic material obtained by a physicochemical process in which silica and petroleum coke, leading to the formation of SiC crystals. SiC can absorb electromagnetic energy and be heated efficiently. It has a dielectric loss factor of 1.71 at 2.45 GHz frequency at room temperature. This ability for microwave absorption is explained by the semi-conductivity of the material [61]. Recently, researchers [56] explored the effect of adding activated carbon, a ceramic additive with good dielectric properties that promotes microwave heating.

4.5 Effect of Healing Cycles on the Asphalt Crack-Healing

Various researchers have presented results on the healing level of asphalt mixtures heated by microwaves. Figure 20a shows results published by Karimi et al. [56] on asphalt mixtures treated with activated carbon. The authors conditioned the circular samples at the testing temperature for two hours before testing. The microwave radiation heated the specimens at 90–100 °C. Once the heating finished, the specimens were kept at room temperature (25 °C) for 24 h for the next loading cycle. Figure 20a shows that the peak load decreases with increasing healing cycles. The peak load at

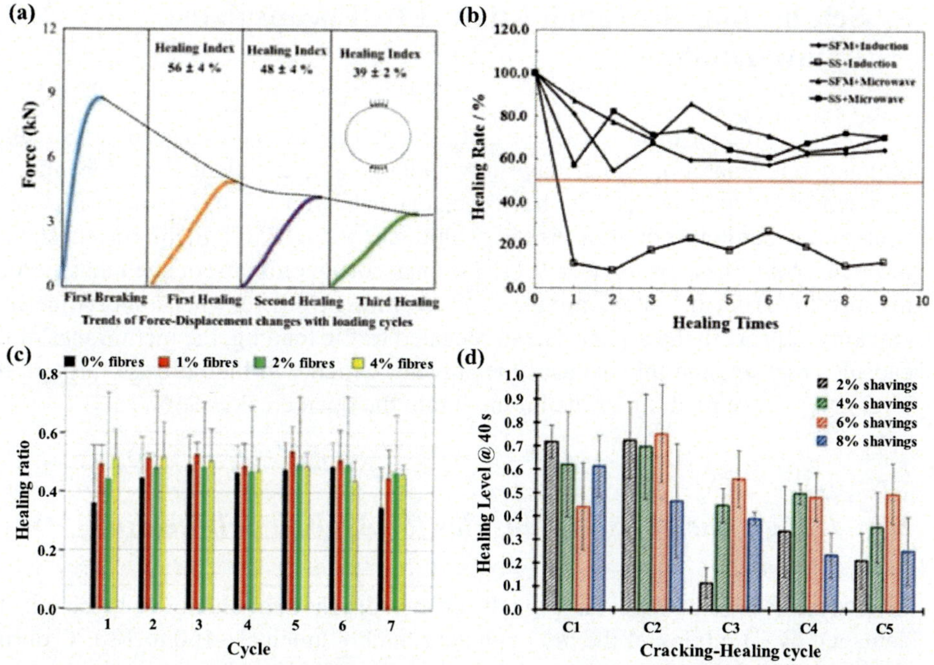

Fig. 20 Results of crack and healing by microwave heating on asphalt specimens prepared with **a** activated carbon (reproduced from Karimi et al. [56]), **b** steel slag and steel fibre (reproduced from Sun et al. [60]), **c** steel wool fibre and reclaimed asphalt pavement (Modified from Gonzalez et al. [49]) and **d** metal shavings (Modified from Norambuena-Contreras et al. [57])

the first breaking cycles was approximately 9 kN, and the healing level at the first, second and third healing cycles was 56%, 48% and 39%, respectively.

Sun et al. [60] obtained a higher healing level of asphalt mixtures heated by microwaves. Additionally, they conducted heating through electromagnetic induction. The researchers concluded that asphalt mixture with steel slag, which was healed by induction heating, showed the worst healing ratio. However, the healing ratio remained between 60 and 85% for all the other type of specimens. Figure 20b shows that healing ratio remains relatively constant after the second healing cycle. Figure 20c shows results published by Gonzalez et al. [49] with different content of metallic fibre and 30% of reclaimed asphalt pavement (RAP). The healing level remained constant within the range of 35–45% for most of the mixtures after the second healing cycle for the different metallic fibre contents. Norambuena-Contreras et al. [57] published results on dense asphalt mixtures with different contents of metal shavings and heated for 40 s (see Fig. 20d). After the second healing cycle, the average healing level decreased significantly, particularly for the mixtures with the lowest content of shavings (2% in volume).

5 Self-healing of Asphalt Mixture by Encapsulated Rejuvenators

To accelerate the intrinsic self-healing of asphalt, oil-based bitumen solvents (called bitumen rejuvenators), which have low molar mass, excellent diffusion into the bitumen and reduce its viscosity have been encapsulated [36, 62]. Examples of rejuvenators are sunflower oil, low viscosity bitumen, or engine oil. In the current state of the technology, the capsules (see Fig. 21) consist of a membrane or shell that protects the solvent during the asphalt mixture manufacturing [63]. So, when a short time after cracking appears in the asphalt due to repeated traffic loading, the membranes of the capsules rupture, and this releases the solvent, which will then diffuse through the bitumen, promoting drain of the bitumen into the open cracks.

5.1 *Design Variables of Capsules for Asphalt Self-healing*

To create effective capsules for asphalt self-healing they should not be mechanically damaged by (i) mixing of the hot aggregates and bitumen at ~160 to 180 °C during manufacture of the asphalt, (ii) hot compaction of the mixture after laying on a road surface at ~140 °C by a roller compactor or (iii) in-service ambient temperature compaction/deformation of the asphalt arising from the action of traffic. They should also (i) not cause sub-optimal road performance, e.g. rutting and reduced skidding resistance and (ii) be cost-effective, i.e. their added cost should not be greater than the benefit obtained from the maintenance reduction.

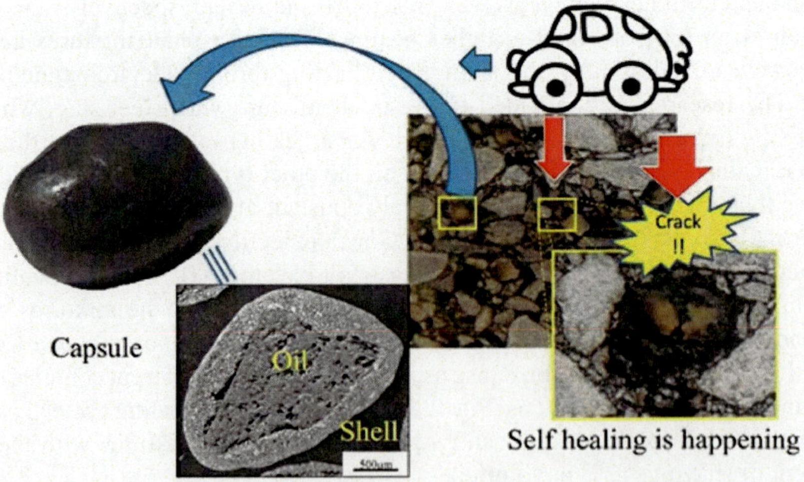

Fig. 21 Working mechanism of self-healing of asphalt mixture by the action of encapsulated rejuvenators containing sunflower oil as a rejuvenating agent (Reproduced from Garcia et al. [62])

The different variables that influence the timely release of the rejuvenators and should be considered for the design of novel capsules for asphalt self-healing are:

1. Location in the road structure.
2. Type of binder used and packing level of the asphalt.
3. Traffic loading level.
4. Environmental moisture and temperature.
5. Size of the capsules.
6. Payload and internal structure of the capsules.
7. Viscosity, affinity of the rejuvenator for the bitumen and aggregates, and molecular weight of the rejuvenator.
8. Mechanical properties and porosity of the membrane/shell.
9. Cost and availability of the capsules.
10. Environmental and health safety.

Based on the list of properties mentioned above, one can understand that in future, a range of encapsulated rejuvenators may be available for different types of roads according to the aggregate gradation and type of bitumen and aggregates used, weather, and road's category.

5.2 Overview of the Existing Capsule Types for Asphalt Self-healing

The first capsules for asphalt self-healing purposes were developed by Garcia et al. [64] (see Fig. 22). These consisting of four parts: an organic or inorganic oil, such as soft bitumen or sunflower oil, porous sand particles of between 1 and 1.7 mm, and a rigid membrane composed of cement and epoxy. The porous sand particles were selected for their ability to absorb liquid [65]. The epoxy-cement membrane was designed so that the capsules could survive the asphalt mixing and compaction process [66] and it was demonstrated that these capsules could release the rejuvenator in a timely manner, which would reduce the stiffness of the asphalt mixture as it was expected [62–66]. Limitations of these capsules were the high cost of the membrane, their low payload, and low efficiency releasing the rejuvenator.

Su et al. [67] encapsulated sunflower oil in a methanol-melamine–formaldehyde shell. The average diameter of the capsules was 10 μm, and they survived the binder mixing and compaction processes; besides, it was demonstrated that the membrane could rupture by the effect of cracks in bitumen and release the sunflower oil, which contributed to improving the drain of bitumen into and closed the cracks [68].

Yang et al. [69] made capsules with an average size of 50 μm, which membrane was composed of nano-$CaCO_3$ particles and methanol-modified melamine–formaldehyde. These capsules showed excellent thermal stability and mechanical properties, which allowed them to survive the melting of bitumen; however, trials in asphalt mixture have not been reported, and their ability to survive the mixing and compaction process or their ability to heal remains unknown.

Fig. 22 Composition of the first porous capsules manufactured for the self-healing of asphalt mixture

Tabaković et al. [70] and Aguirre et al. [71] developed calcium-alginate and sodium-alginate hollow-fibres, respectively, which contained sunflower oil. The fibres had a diameter of approximately 100 μm and could be cut to the desired length. They consisted of a string of individual oil compartments protected by a calcium- or sodium-alginate membrane. It was found that the hollow-fibres could efficiently improve the drainage of bitumen into the cracks, so as the stiffness of dense asphalt. On the other hand, the fibres did not improve the self-healing when the asphalt was subjected to fatigue damage [72] and, the reasons for this remain unclear.

Al-Mansoori et al. [63] and Norambuena-Contreras et al. [73] encapsulated sunflower oil in a spherical multi-core calcium-alginate matrix, which had an average size of 2.5 mm (see Fig. 23a). These capsules were added as an additive during asphalt mixer (see Fig. 23b), and it was found that they distributed homogeneously in the asphalt mixture and only approximately 1% of capsules broke during asphalt mixing and compaction (see Fig. 23c) [73].

Additionally, the capsules did not affect the rutting, fatigue life, particle loss, or skidding resistance of asphalt mixture, improved the crack self-healing properties of asphalt and contributed to extend its fatigue life when the fatigue test was done including resting periods. In reference [74], it was found that the amount of these capsules that did not affect the usability and performance of asphalt that contains capsules was 0.5% by total weight of the mixture. These results could be reproduced in dense and porous asphalt mixtures, as well as stone mastic asphalts, see

Fig. 23 Capsules for the self-healing of dense asphalt mixture: **a** Multi-core calcium-alginate capsules; **b** Mixing of capsules in hot dense asphalt mixture; **c** 3D reconstruction of the capsule spatial distribution inside of the asphalt core (Modified from Norambuena-Contreras et al. [73])

Norambuena-Contreras et al. [75]. However, the main disadvantage of using calcium-alginate as the membrane of the capsules for asphalt self-healing is that it allows the diffusion of sunflower oil in times that range from weeks to months.

5.3 *Characterization of the Asphalt Self-healing Promoted by Encapsulated Rejuvenators*

One of the main challenges for the development of encapsulated rejuvenators is the characterization of asphalt self-healing that is promoted by the capsules. In addition to the continuous passage of vehicles, materials in the road surface are subjected to ageing and chemical changes due to oxidation, thermal cycles and moisture, which are complex to reproduce in the laboratory. Furthermore, it is not known yet if the capsules break due to cracks in the asphalt or to accumulated cyclic damage that also damages and cracks the asphalt. To date, three methodologies that aim to reproduce the roads conditions exist and these are described below.

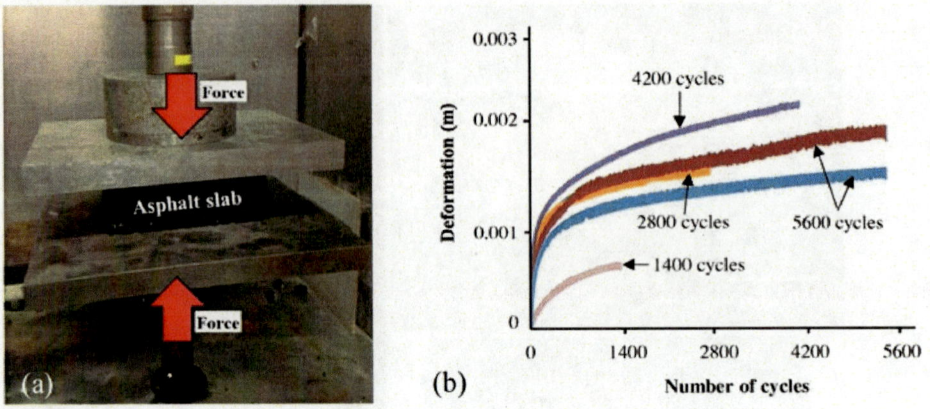

Fig. 24 Cyclic compressive loading test: **a** Experimental set-up; and **b** Deformation versus the number of cycles of asphalt slabs with capsules (Reproduced from Garcia et al. [62])

5.3.1 Unconfined Cyclic Compressive Loading Test

This method was developed by Garcia et al. [62]. Its objective is to simulate the effect of multiple passing vehicles on the rupture of the capsule's membrane and characterize the amount of oil released from the capsules. It uses an asphalt slab, and it is done by applying an unconfined cyclic compression load on the top face of the slab (see Fig. 24a). Furthermore, the test specimens can be cylindrical, square, or prismatic, as soon as they have two flat faces at the top and the bottom. Additionally, the amount of cycles that simulate several years of traffic loading (see Fig. 24b) can be related to the loading level of the test by means of fatigue curves, which must be estimated for each type of pavement material being considered.

The main disadvantage of this test method is that, because it does not produce a fracture of the asphalt, the self-healing cannot be easily quantified; on the other hand, this method is excellent to characterize the breaking and release behaviour of capsules in asphalt mixture. For example, in the experiment which results are reflected in Fig. 25a, a constant cyclic load was applied on the asphalt test samples with and without capsules [62]; and the vertical axis reflects the amplitude of each loading cycle; it can be observed how the asphalt with capsules always had higher amplitude, which grew with the number of cycles. It was assumed that this reflected the rupture of the capsule's membrane, as it is shown in Fig. 25b.

5.3.2 Crack-Healing and Confined Static Compressive Loading Test

This crack-healing test was reported in the first place by Al-Mansoori et al. [74] (see Fig. 26), and it is applied to asphalt prismatic test specimens with and without capsules that have been previously cracked under three-point bending test (see Step

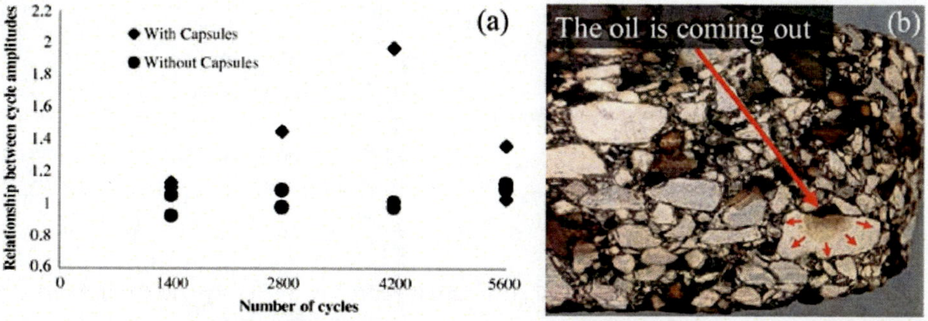

Fig. 25 Results of the compressive loading test: **a** Relationship between the average cycle amplitudes before and after resting period versus number of loading cycles, for samples with and without capsules; **b** Test sample after testing, with capsules broken (Modified from Garcia et al. [62])

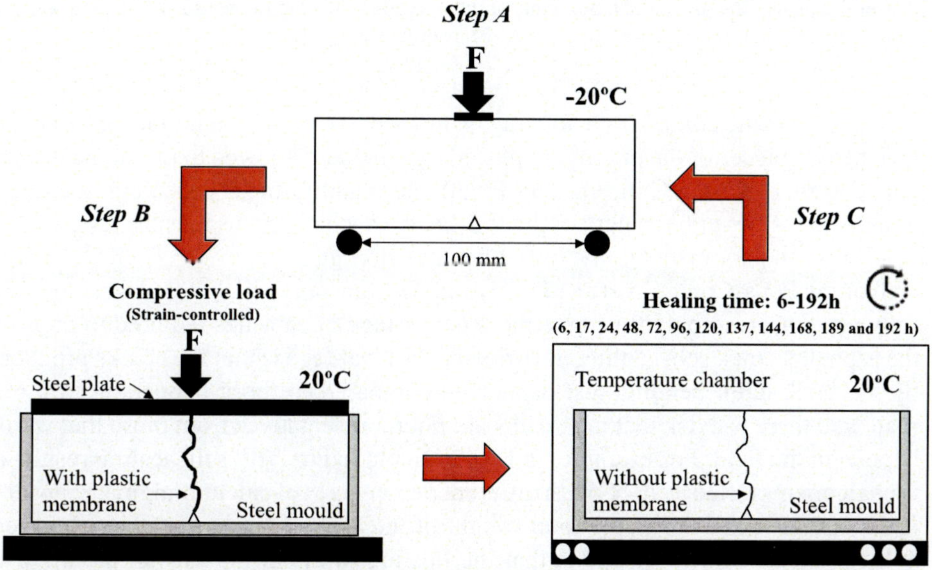

Fig. 26 Schematic diagram of the asphalt mixture crack-healing and confined static compressive loading test: Step A. Crack generation in the asphalt beam; Step B. Capsules activation to begin the oil release; and Step C. Healing process (Reproduced from Al-Mansoori et al. [76])

A in Fig. 26). In order to avoid creeping and permanent deformation of the asphalt, the asphalt prisms are conditioned at −20 °C before the three-point bending test.

After the asphalt prisms are introduced in a steel mould that prevents lateral deformation and a confined static compressive load is applied to the surface of prisms, to simulate traffic loading and break the capsules (see Fig. 26). To avoid the drain of bitumen into the crack during the cyclic loading, a shape-adaptable membrane is placed in the crack (see Step B in Fig. 26).

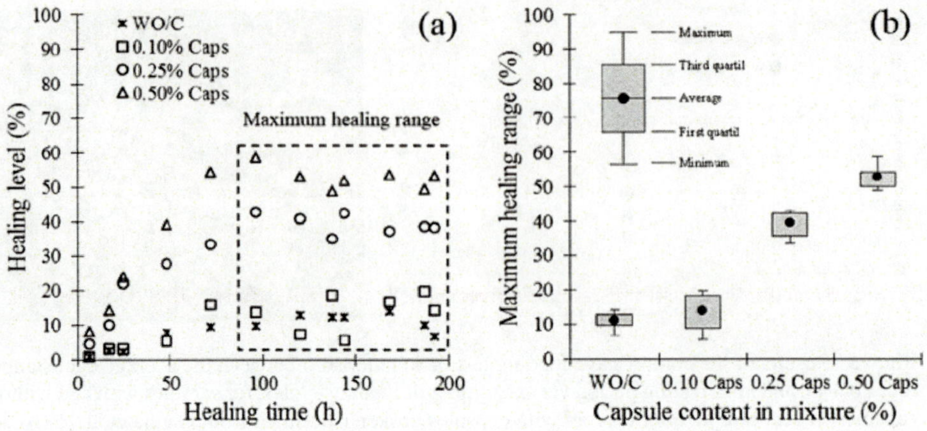

Fig. 27 Self-healing results for asphalt mixture samples with and without capsules: **a** healing level depending on healing time; and **b** box plots of the maximum healing range (96–196 h) with capsule content in the mixture (Reproduced from Al-Mansoori et al. [76])

Once the static compressive loading is finished, the plastic membrane is removed, and the two pieces of the prisms are placed together and allowed to rest at the desired temperature (see Step C in Fig. 26). Finally, to quantify the self-healing, the three-point bending flexural strength of the self-healed asphalt prisms must be tested under at −20 °C and compared to the initial flexural strength.

Norambuena-Contreras et al. [73, 75] and Al-Mansoori et al. [74, 76] have successfully used this test to optimize the mass percentage of capsules required to improve the asphalt's crack self-healing. Its primary disadvantage is that the cracks produced by the three-point bending test cannot be compared to those happening in a real road, and there is a risk that the results are not representatives from those that would happen in the field. For example, in Fig. 27, included in [76], self-healing results of asphalt prisms tested at 20 °C with different percentages of calcium-alginate capsules (0.10%, 0.25% and 0.50% by total weight of the asphalt mixture) can be observed. What is evident from Fig. 27 is that the healing levels in the asphalt prisms with capsules were higher than in prisms without capsules and that the healing level of asphalt mixtures with and without capsules increased with the healing time until a maximum value, and then remained constant (see Fig. 27a).

This result is reasonable due to the fact that a higher capsule content increases the probability of breaking more capsules when the mixture is subjected to external damage (see Fig. 27b), which increases the potential released oil content in the mixture [74].

In conclusion, this methodology can be used to characterize the effectivity of capsules for asphalt self-healing and to compare different types of capsules but it does not quantify the increase of durability of an asphalt road by the effect of the rupture of the capsules.

5.3.3 Cyclic Wheel-Track Test Over an Elastic Foundation

The objective of this test is to realistically simulate a bottom-up crack happening as a result of a soft spot at the lower layers of a pavement (see Fig. 28). At the time of writing this section, this test is under development at the Nottingham Transportation Engineering Centre (NTEC) from the University of Nottingham, UK. To perform the test, an asphalt prism that contains capsules is simply placed over an elastic foundation and a rolling wheel passes and applies a cyclic load over the asphalt sample (see Fig. 28a).

This cyclic test produces multiple micro-cracks in the middle of the asphalt prism, and rupture the capsules, which causes the rejuvenator's release. To simulate cycles with lower traffic intensity, such as day/night cycles, resting periods are simply applied during the cyclic loading; during these periods where cracks are not appearing or growing, the bitumen can drain into and fill the cracks. One advantage of this method is that the loading level can be adjusted to represent the traffic over an asphalt road; furthermore, the resting periods can be implemented to represent different circumstances in the road.

Examples of the results produced by this test method are shown in Fig. 28b. The rest period was a period of 24 h, at 20 °C at the number of cycles indicated in the horizontal axis; afterwards, more loading cycles were applied until the prism's breakage. The healing index represented in the vertical axis is a function of the additional number of cycles that the asphalt prisms resist thanks to the rest periods. In this case, one means that the durability has been doubled, while 0 means that the durability remains constant. As a result of this testing method, it can be observed that there is a moment when the capsules are most effective, which probably corresponds to the time when they break; furthermore, this methodology can be used to predict the increase of durability that road will have.

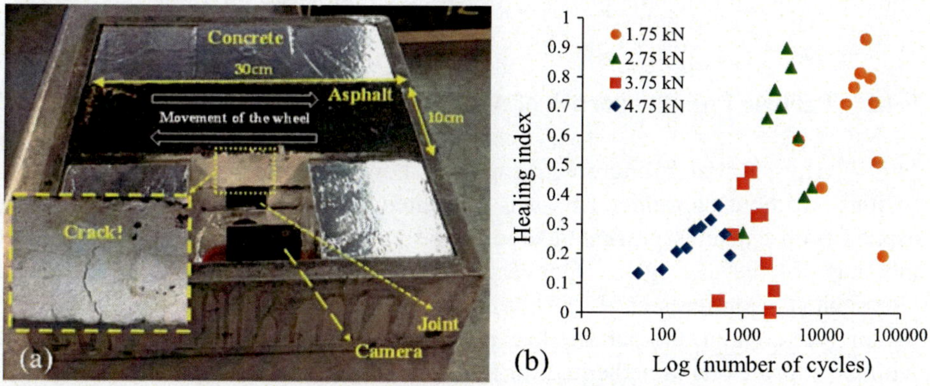

Fig. 28 Cyclic wheel-tracking test over an elastic foundation: **a** Experimental set-up; **b** Results of healing index at different number of cycles (Unpublished images)

Finally, by comparing the number of cycles without rest periods resisted by the asphalt with and without capsules, it can be determined if the capsules have an appropriate resistance for the road, or if they will promote cracking.

5.4 Usability and Performance of Road Surface Materials that Contain Capsules

To promote self-healing, capsules should not reduce the durability of asphalt roads, decrease their skidding resistance, or increase the roughness by promoting rutting or potholing. Logically, to avoid quality reductions of the asphalt surfaces, the capsules should have surface energy equivalent to that of aggregates, be non-porous and, as stiff and resistant as an aggregate to avoid affecting the solid skeleton structure of asphalt mixture. To date, very few published studies have focused on these properties, and most of them have focused on dense asphalt mixture and stone mastic asphalt (SMA); this last is a mixture with an open gradation, where the voids have been filled with mastic. The existing results are summarized below:

5.4.1 Physical Properties of Asphalt Mixture with Capsules

In [74], it was reported that 0.5% of capsules by weight of total mixture increased the density and reduced the air void content of dense asphalt mixture. The reason for this was that the capsules used could adapt their shape to that of the aggregates during compaction. On the other hand, in [75] stiffer capsules were used, and it was reported that 0.5% of these capsules by weight of total mixture did not affect the bulk density or air void content of SMA, which proved that a capsule content of 0.5% did not adversely influence the physical properties of the SMA mixtures.

5.4.2 Particle Loss Properties of Asphalt Mixture with Capsules

In [74], dense asphalt mixture with capsules showed lower particle loss than asphalt mixtures without capsules. This was attributed to the lower air void content that asphalt with capsules presented, due to the capacity of the capsules to adapt to the shape of the aggregates. Conversely, in an experimental study developed by Norambuena-Contreras et al. [77], where capsules were stiffer and did not reduce the air void content, the particle loss resistance of dense asphalt mixture with, and without, capsules remained equal for asphalt mixtures containing 0.5% of capsules.

5.4.3 Effect of Capsules on the Asphalt Stiffness and Rutting Properties

In [74] and [75], who added 0.5% of calcium-alginate capsules to dense asphalt and SMA mixtures, respectively, it was found that the average values of stiffness modulus were similar for specimens with, and without, capsules regardless of the test temperature. Furthermore, in [78], it was found that the addition of 5% of calcium-alginate fibres increased the stiffness of the dense asphalt mixture in between 30 and 45%. In addition, it is known that there is a relationship between the stiffness of the asphalt and its rutting properties. Based on this, in [74], it was concluded that the capsules did not increase the risk of rutting in asphalt roads.

5.4.4 Effect of Capsules on the Skidding Resistance of Asphalt Roads

In [75], it was reported that the rejuvenators in the capsules could reduce the skid resistance of SMA surfaces by approximately 7%. However, considering the variation of the results for the tested mixtures with, and without, capsules this reduction was not significant. Therefore, it was concluded that the oil released by the capsules did not increase the risk of skid on the surface of the SMA asphalt pavement.

5.4.5 Fatigue Life Properties of Asphalt Mixture Without Resting Periods

In [74], it was found that the fatigue life of asphalt mixture with capsules was reduced by approximately 15%. This was attributed to softer capsules that ruptured before and promoted cracking in the asphalt mixture. On the other hand, in the study developed by Norambuena-Contreras et al. [77], where stiffer calcium-alginate capsules were used, it was found that the fatigue durability of dense asphalt was not affected and it can be even increased by the effect of the capsules.

5.5 Design of the Mechanical Properties of Capsules

Mechanical characterization is essential to understand the capsules behaviour during their manufacturing process and on-site use. The mechanical characterization of capsules also allows us to know several parameters, such as elastic modulus and failure/strain, which are vital factors to determine the final applications [79].

Furthermore, understanding the parameters to characterize the mechanical properties of capsules is of excruciating importance to design capsules that do not break too early or too late during the lifetime of asphalt roads. As it was mentioned above, the number of parameters to design capsules is so high that their mechanical properties can be considered only numerically.

The two main challenges to model the interaction of capsules with the asphalt are: (i) the solid–fluid interactions between the core and the membrane and (ii) the high deformations and fracture of the membrane. There are previous scientific studies based on the finite element method (FEM), which modelled the fluid–solid interactions and the solids deformation, such as Walter et al. [80], where the interaction of a spherical microcapsule and the external liquid flow was analysed by coupling FEM and a boundary integral method to solve the Stokes flows. Although large deformations were accurately modelled, the fracture of solids is a pending issue for this type of approach. The solid–fluid interaction is another drawback of FEM, for example in Müller et al. [81] a method to simulate fluid–solid interactions based on coupling smoothed particle hydrodynamics (SPH) and FEM was presented, which is not able to reproduce the fracture of solid particles.

Recently, a mesh-free discrete multi-physics (DMP) approach has been implemented to validate the interactions between fluid and solid particles [82, 83]. DMP is based on a particle framework which includes SPH to simulate the hydrodynamics of liquid, and the mass and spring model (MSM) to simulate the mechanical properties of the solid and to simulate high deformations and stable fracture. The SPH model and MSM share the common particle-based paradigm and the same flowchart [84] (see Fig. 29). Although each one is focused on diverse types of internal forces, (SPH on hydrodynamics and MSM solid deformation), the algorithm is very similar in both cases. All the particles from both models are considered as Lagrangian nodes (interpolation method), and at the same time, their position is updated by Eq. 3:

$$m_i \frac{\mathrm{d}v_i}{\mathrm{d}t} = m_i \frac{\mathrm{d}^2 r_i}{\mathrm{d}t^2} = \sum_{i \neq j} F_{i,j} + \sum F_E \tag{3}$$

where m is the mass, v the velocity, r the position, F_E the external forces and $F_{i,j}$ the internal forces. Equation 1 is the basis to link both models by means of the mathematical term of internal forces ($F_{i,j}$).

The SPH equations of motion come from the discretisation of the Navier–Stokes equation [85] where each particle has an associated mass, velocity, pressure and density obtaining the Eq. 4:

$$m_i \frac{\mathrm{d}v_i}{\mathrm{d}t} = \sum_j m_i m_j \left(\frac{P_i}{\rho_i^2} + \frac{P_j}{\rho_j^2} + \sqcap_{i,j} \right) \nabla_j W_{i,j} + \sum F_E \tag{4}$$

where P is the pressure, m is the mass, ρ is the density and v is the velocity, for each particle. Furthermore, W means the Kernel parameter and $\sqcap$ introduces the viscous forces.

MSM is a model where the motion and the interaction between particles are studied. This model is based on Newtonian equations of motion, as follows:

$$m_i \frac{\mathrm{d}^2 r_i}{\mathrm{d}t^2} = -\frac{\partial}{\partial r} U_{\mathrm{tot}}(r_1, r_2, \ldots r_N) + \sum F_E \tag{5}$$

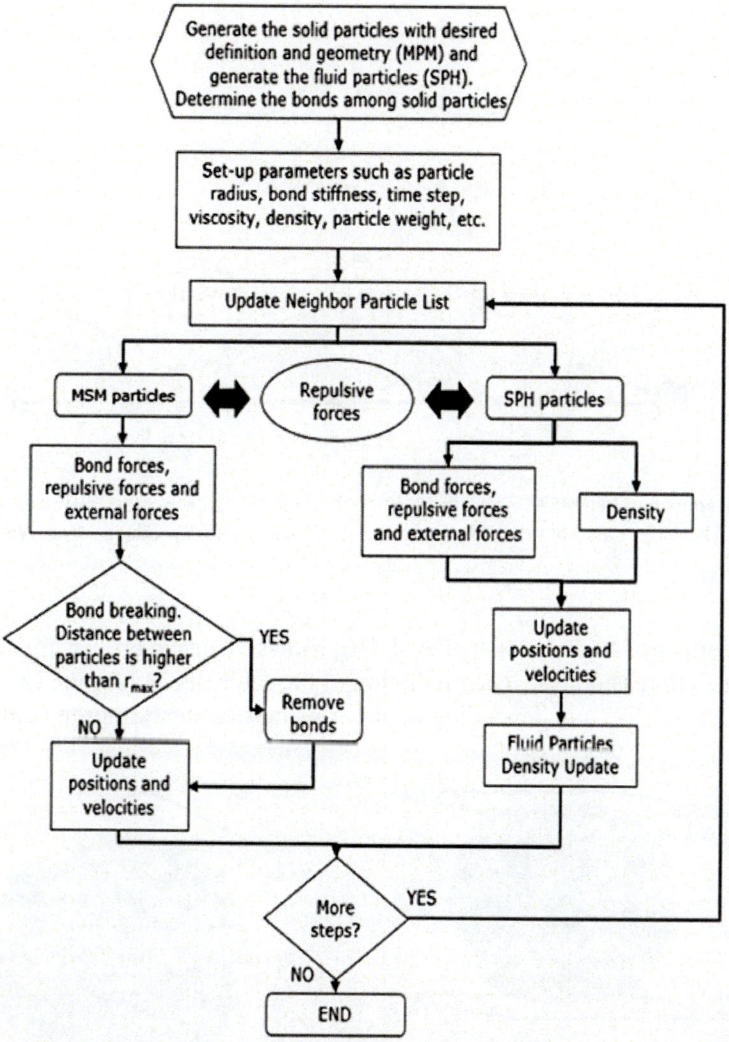

Fig. 29 Computational flowchart for coupling MSM and SPH (Modified from Wu et al. [84])

The internal forces term changes among the different models. In this case, U_{tot} is the total interatomic potential and it is divided into non-bonded and bonded. The non-bonded interactions are computed on the basis of the neighbour list (a list of non-bonded atoms within a certain radius; see Fig. 30). The bonded forces can be represented for many different types of potentials.

DMP can also model the breakage of solids. As the algorithm calculates the distances between particles for each time step, the breakage can be defined as a condition. When the distance between the particles is higher than a determined value, the computational spring breaks starting the fracture. This function has many applications for simulating solid fractures caused by loads of different origins. For example,

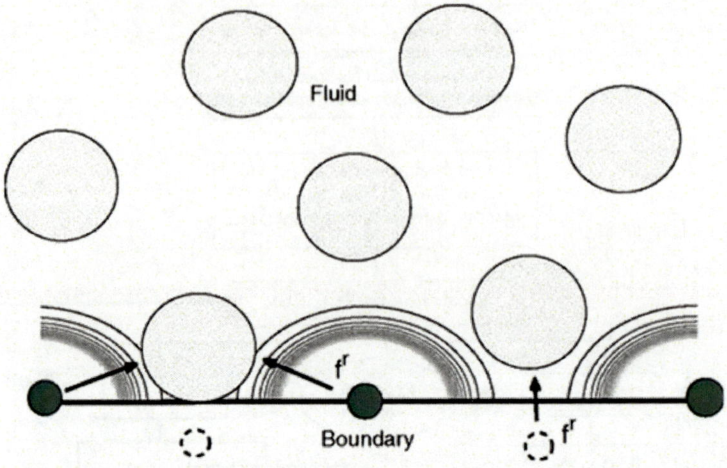

Fig. 30 Repulsive SPH-MSM contacts when the fluid particles reach a certain distance to the MSM particles. This effect also is reached among the fluid particles (Reproduced from Van Liedekerke et al. [86])

Fig. 31 contained Van Liedekerke et al. [86] shows a capsule broken after the simulation of a compression test. The prediction of the force needed for the capsule failure and the type of fracture depending on the capsule structure are promising outcomes

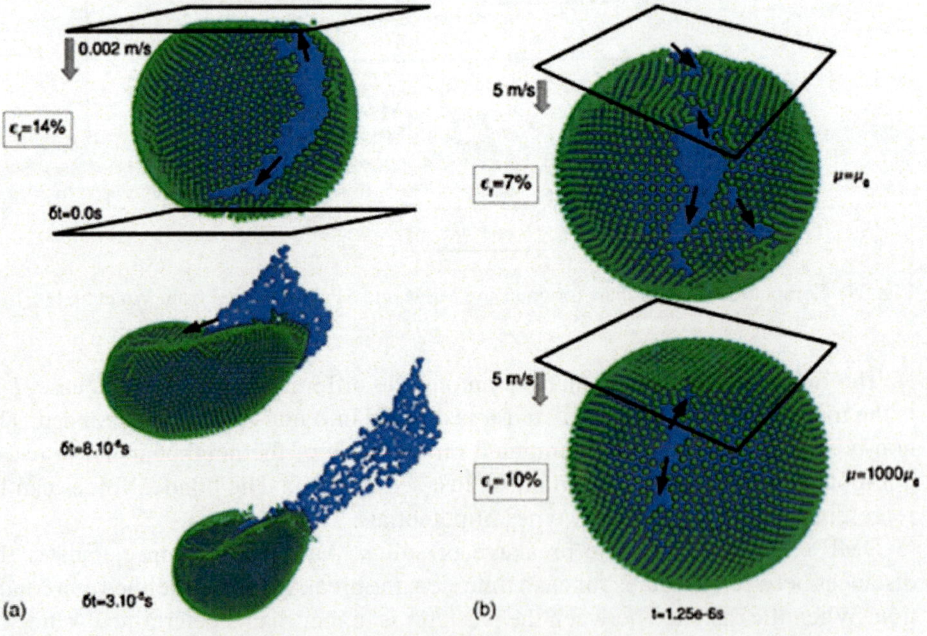

Fig. 31 Snapshots of the computational simulation fracture for capsules with different strength values after a quasistatic compression test (Reproduced from Van Liedekerke et al. [86])

for the application of the particle-based multi-physics approach. The capsules shown in Fig. 31 are examples of capsules with different strengths [86]. It is also possible to simulate different rigidity of the membrane and viscosity of the core, to simulate novel capsules with desired rupture properties.

6 Large-Scale Application of the Induction Healing in Asphalt Roads: The Study Case of the HEALROAD Project

As discussed in this Chapter, induction heating is a preventive and non-intrusive maintenance technique used to accelerate the self-healing properties of asphalt mixture with metallic fibres, allowing it to extend the lifespan of the road pavements.

During 2015–2018 years, HEALROAD project (induction heating technique to increase road durability and reduce maintenance costs and disruptions) funded by the European Programme Era-Net Plus Infravation was developed with the aim of underpinning the industrialization of this asphalt healing technology [87]. Partners of the project were the Universities of Nottingham and Cantabria, SGS Intron B.V., Bundesanstalt für Straßenwesen (BAST) and Heijmans B.V. These partners worked for 30 months with an approximated budget of 1.2 M€.

The HEALROAD project involved a first laboratory stage to understand the fibre-reinforced asphalt mixture properties affecting their induction healing capacity. The end of the project, the induction heating technology validation was carried out through a large-scale demonstrator [88].

With these purposes, a full-scale size induction heating machine was optimized to increase the temperature of an inductive asphalt mixture was designed. Besides, the accelerated pavement testing (APT) facilities available at BAST (duraBAST) was used to simulate several years of traffic in just a few days [89], to select the equivalent number of axle loads that are required before using the induction heating machine in a road subjected to real traffic [90]. The following section provides information on the construction of a test section demonstrator and the APT investigation program carry out within the framework of the Infravation HEALROAD research project.

6.1 Test Section Construction and Large-Scale Research Program

The HEALROAD test section was built in the experimentation area of the Federal Highway Research Institute (duraBASt) in Germany; see Fig. 32a. The area is located on the Cologne east motorway junction (BAB A3/A4). The entire facility covers an area of around 25,000 m^2 and has a total length of 1 km.

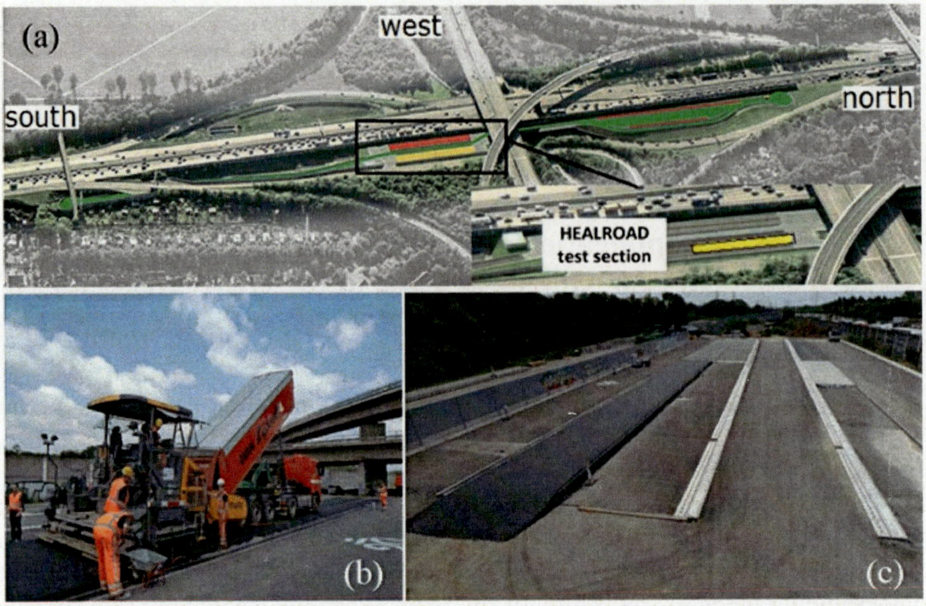

Fig. 32 **a** Overall view of the demonstration, investigation and reference area of the Federal Highway Research Institute (duraBASt) facility, and the HEALROAD test section position; **b** construction process of the HEALROAD test section and; **c** view of the test section after the end of construction phase (Images provided by Bastian Wacker)

The pavement design for the test section was selected based on the criteria to have a robust supporting structure for the HEALROAD surface layer. Four layers were constructed over the prepared sub-grade to get a complete structure of 40 cm. The static load plate test (PDV) and the lightweight deflectometer (LFG) methods were used for the evaluation process of the test section.

The construction indicates three types of asphalt layers. The base layer and binder layer were designed based on German guidelines. The asphalt mixture containing 0.45% vol. /mixture of steel wool fibres used for the HEALROAD test section was produced at a Heijmans' standard production plant in the Netherlands and transported to duraBASt to be paved by standard paving equipment (Fig. 32b, c). The project considered that one of the most common failure mechanisms in surface layers is the ravelling (particles loss) due to a combination of shear and normal stress with micro-fatigue processes in the bounding phase of the mixture.

With the objective of simulating the micro-fatigue phenomenon in a short period of time, the BASt Mobile Load Simulator MLS30 was used (Fig. 33a) applying 150,000 loading cycles in each loading area by a single wheel (75 kN) at a linear speed of 7 km/h (2000 cycles/h) and a transverse oscillatory shifting from centre to 350 mm at each side at 4 mm/s (50 mm transverse moving in each step).

Otherwise, the induction heating treatment on the HEALROAD test section was undertaken by a full-scale induction machine developed by SGS INTRON; see Fig. 33b. This machine consists on a train of two rectangular induction coils with a

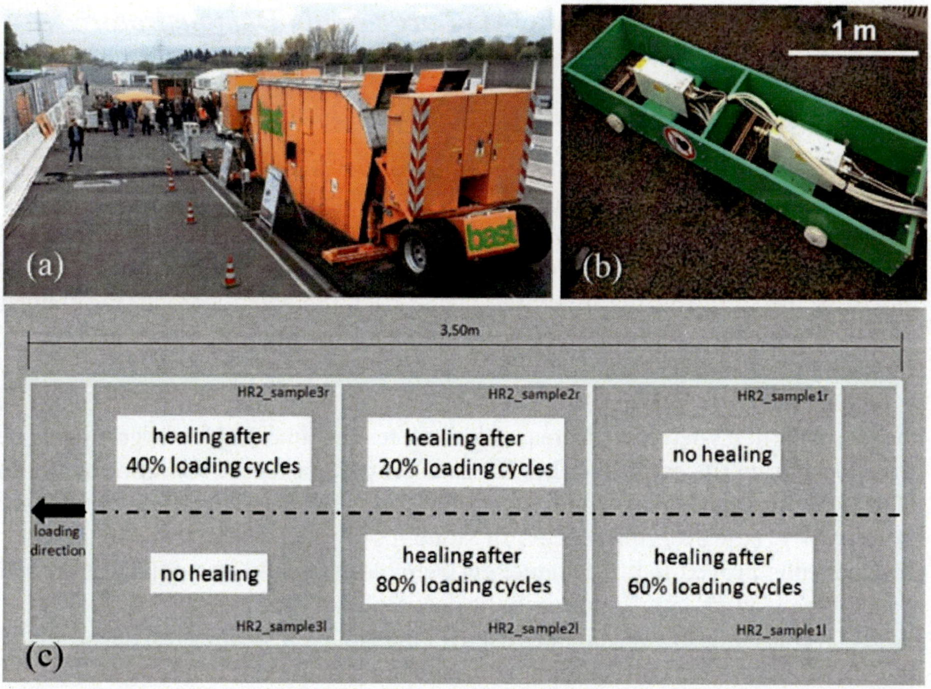

Fig. 33 **a** Mobile Load Simulator MLS30 working on the HEALROAD test section; **b** full-scale induction machine developed by SGS INTRON; and **c** second loading area with two reference and four healing samples (Images provided by Bastian Wacker)

width of 0.4 m connected to two independent generators with a potential of 10 kW, one cooling system with capacity for 300 l, one power unit of 35 kVA and a self-driving machine with steering wheel and eight wheels on the front axle (compacting simulation and force spreading) and four wheels on the back axle [88].

By means of the Simulator MLS30 and the induction machine, six loading areas were tested under large-scale conditions (see Fig. 33c). Two of them were considered as control sections (i.e. no healing treatments were applied on them). The other four sections were healed after 20%, 40%, 60% or 80% of the total APT loading cycles. To assess the ravelling and the healing effect on the HEALROAD test section, the road surface was continuously inspected by image processing, and any particle detached was collected and registered for accounting the damage [88].

Finally, to compare the loss particles between large-scale testing and laboratory conditions, several rotating surface abrasion tests (RSAT) on different asphalt drill cores were performed at various testing times. Standard RSATs were developed on cores after the construction and after loading and induction heating drill cores also were used to evaluate the effect of the induction process. The cores after loading were taken from the different areas of the loaded area, as shown in Fig. 33c.

6.2 Results of the HEALROAD Large-Scale Demonstrator

The accelerated pavement testing (APT) by MLS30 was successfully performed on the HEALROAD test section. In addition, various measurement methods (including bearing capacity, density measurements and stone loss) showed that the loading influenced the pavement structure. However, the first large-scale APT was not as representative as hoped because the influence of the induction healing effect could not be conclusively demonstrated in the short program. Two points are decisive for this, first, the short loading time used in the APT trails and, second, the standard speeds of the lateral movement with 4 mm/s. For this preseason, the possibility to create micro-cracks was limited by the generated shear forces.

During the testing of the HEALROAD full-scale demonstrator, it was found that the size of particles detached decreases over the testing time. Hence, larger aggregates were lost during the early stages, while smaller particles were lost towards the end of the tests. In contrast, the stone loss results during the RSAT were ten times higher than in APT. However, the process of stone loss [g/h] in APT is higher. In short, RSAT results proved that the induction technology should be applied at an early stage of the loading process to reach the best asphalt healing results.

A unit consisting of two coils was used for the induction heating test on the HEALROAD full-scale demonstrator. From infrared images results, it was observed that the first coil produced the first excitation before the second coil caused the next heating. As can be seen in Fig. 34, the induction energy achieved very homogeneous heating of the surface layer, which is also visible approx. Ten minutes after the measure (see right picture). However, during the research program, the settings were adapted so that the target temperature of 100 °C could be reached during the second crossing. It was concluded that this result depends mainly on the speed of the induction machine, but also on the settings of the induction devices.

In summary, a full-scale test was suitable to prove that HEALROAD mixtures can be manufactured in conventional asphalt plants and laid-down and also compacted by conventional methods. Also, the temperature of the road surface was homogenous during induction tests, which showed that no fibre clusters were produced. Finally, visual inspections found no metal spikes on the road surface.

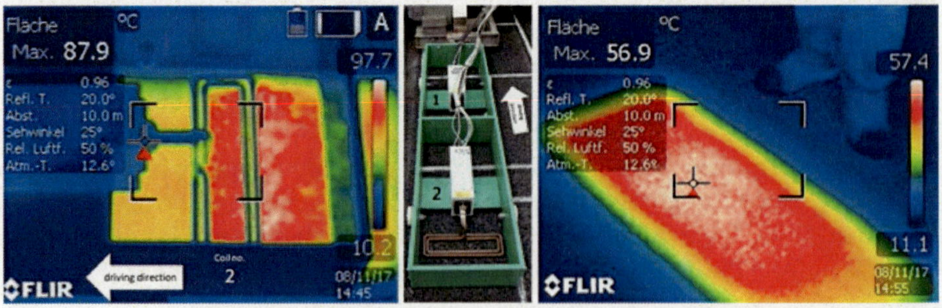

Fig. 34 Results of infrared images after induction heating test on the HEALROAD test section

7 Summary and Outlook

The present chapter provided a comprehensive review of the most recent advances on self-healing of bituminous materials (i.e. asphalt mixtures, mastics and bituminous binders) from the engineering technologies associated to promote the crack-healing capability in asphalt pavements to the large-scale applications.

Currently, the acceleration of the naturally self-healing property of bitumen has been investigated by two main approaches. First, an extrinsic self-healing approach to reduce the viscosity of bitumen by increasing its temperature through by externally triggered heating using electromagnetic induction heating and microwave radiation and, second, an intrinsic self-healing approach that reduces the viscosity of bitumen by the effect of encapsulated rejuvenating agents embedded inside the asphalt mixture.

To perform asphalt self-healing by induction heating, the addition of highly conductive materials such as steel fibres is needed. The bituminous material is exposed to a high-frequency alternating electromagnetic field (in the order of kHz), which induces eddy currents in materials that are electrically and magnetically susceptible. So, metallic fibres are heated by the induced eddy currents, and the heat energy diffuses into the bitumen. Once the temperature of bitumen is above its softening point, this material will flow through the micro-cracks, sealing them.

Morphology and distribution of metallic fibres inside the mixture, as well as the temperature reached by the asphalt mixture during the induction heating, are the most important factors concerning the healing rates of asphalt pavements. Overall, short fibres have a good distribution during the mixing process, resulting in better mechanical and healing properties than long fibres, which are prone to form clusters, resulting in low mechanical properties. Additionally, healing and fatigue life extension rates have proved to be in direct relation with the temperature reached by the asphalt mixture during induction heating.

Currently, the addition of novel materials has been explored for induction heating approach in bituminous materials. For example, nanomodifiers such as carbon nanotubes and nanoclays are among the most used and promising nanomaterials for promoting healing of bituminous materials. Nevertheless, critical aspects such as heterogeneous nanoparticles distribution, overheating of bitumen, lack of understanding of the healing mechanisms, large-scale performance, among others, still need significant work.

Moreover, extrinsic self-healing by microwave radiation heating consists on the capability of an electromagnetic field (in the order of GHz) to polarize the charges in the heated material and the inability of this polarization to follow extremely rapid reversals of the electric field produced by microwaves. Asphalt mixtures containing steel wool fibres have also been explored for microwave heating because their capacity to absorb and conduct more thermal energy than bitumen and aggregates. Later, metallic fibres transfer the heat to the bitumen and aggregates, reducing the bitumen viscosity and repairing open cracks.

Additionally, the reduction in mechanical properties due to fibre clustering phenomena has led to consider new materials. In this sense, steel shaving, metallic powders and aggregates, as well as ceramic additives with good dielectric properties such as silicon carbide and activated carbon have been used to improve the heat transfer to the bitumen, resulting in promising results for various healing cycles.

On the other hand, intrinsic asphalt self-healing by the effect of encapsulated rejuvenating agents is considered a revolutionary technology for the crack-healing of aged asphalt pavements. Usually, the capsules consist of a membrane or shell that protects the solvent during the asphalt mixture manufacturing. Therefore, when a short time after cracking appears in the asphalt due to repeated traffic loading, the membranes of the capsules rupture, and this releases the solvent, which will then diffuse through the bitumen, promoting drain of the bitumen into the open cracks.

In the current state, different encapsulation methods have been proposed. Porous sand containing the rejuvenating agent and covered by an epoxy-cement membrane, core–shell capsules, hollow-fibres and multi-core capsules have proved to release the rejuvenating agent into the cracked bitumen effectively.

Besides, since there is still unclear if the capsules break due to cracks in the asphalt or to accumulated cyclic damage, different tests have been carried out in asphalt samples containing capsules under laboratory conditions. Different tests, such as unconfined cyclic compressive loading test, confined static compressive test and cyclic wheel-track over elastic foundation, have proved to be useful in the characterization and quantification of the healing.

Overall, current research has stated that incorporating capsules has no significative effects on the physical–mechanical properties of asphalt pavements, indicating that this technology could effectively heal the cracked asphalt pavement without affecting the physical–mechanical performance. However, more research is needed to characterize the capsules' effect on the different types of asphalt mixtures.

Moreover, the mechanical characterization of capsules is essential to understand the capsules' behaviour during their manufacturing and on-site use. Nevertheless, since capsules' design implies many parameters, numerical modelling based on Discrete Multi-Physics approach proportionate a good approximation for understanding some complex phenomena such as the solid–fluid interactions between the core and the membrane, and the mechanical behaviour of the membrane.

Finally, large-scale applications of self-healing asphalt pavements have been recently carried out. Mainly, the HEALROAD project (2015–2018) was developed with the aim of underpinning the industrialization of induction heating asphalt healing technology. Thus, a full-scale size induction heating machine was designed to increase the temperature of an inductive asphalt mixture. Accelerated pavement testing (APT) was used to simulate several years of traffic. Main results proved that HEALROAD mixtures can be manufactured in conventional asphalt plants and laid-down and also compacted by conventional methods. Besides, the temperature of the road surface was homogenous during induction tests, proving that no clusters of fibres were produced. Overall, the induction technology should be applied at an early stage of the loading process in order to reach the best asphalt healing results.

Acknowledgements Jose Norambuena-Contreras want to thank the funding given by the National Research and Development Agency (ANID) from the Chilean Ministry of Science, Technology, Knowledge and Innovation, through the Research Project FONDECYT Regular 2019 No. 1190027. Nilo Ruiz-Riancho thanks the project H2020-MSCA-ITN-2016 which has received funding from the EU's H2020 Programme for research, technological development and demonstration under grant agreement number 721493. Bastian Wacker would like to acknowledge the Infravation Project (an ERA-NET Plus on Infrastructure Innovation) under the grant agreement no. 31109806.0003—HEALROAD. Jose L. Concha wishes to thank the financial support given by the University of Bío-Bío for his internal PhD scholarship granted.

References

1. Read J, Whiteoak D (2003) The shell bitumen handbook, 5th ed. Thomas Telford Ltd
2. Harmelink D, Shuler S, Aschenbrener T (2008) Top-down cracking in asphalt pavements: causes, effects, and cures. J Transp Eng 134(1)
3. Garcia A (2012) Self-healing of open cracks in asphalt mastic. Fuel 93:264–272
4. Qiu J (2012) Self healing of asphalt mixtures. PhD thesis. Delft University of Technology, The Netherlands
5. Liu Q, Schlangen E, van de Ven M (2013) Induction healing of porous asphalt concrete beams on an elastic foundation. J Mater Civ Eng 25(7):880–885
6. García A, Schlangen E, van de Ven M, Liu Q (2009) Electrical conductivity of asphalt mortar with conductive fibers and fillers. Constr Build Mater 23:3175–3181
7. García A, Schlangen E, van de Ven M, van Vliet D (2010) Induction heating of mastic containing conductive fibers and fillers. Mater Struct 44:499–508
8. Liu Q, Schlangen E, van de Ven M, García A (2010) Induction healing of electrically conductive porous asphalt concrete. Constr Build Mater 24:1207–1213
9. Dai Q, Wang Z, Hasan M (2013) Investigation of induction healing effects on electrically conductive asphalt mastic and asphalt concrete beams through fracture-healing tests. Constr Build Mater 49:729–737
10. Tang J, Liu Q, Wu S, Ye Q, Sun Y (2016) Investigation of the optimal self-healing temperatures and healing time of asphalt binders. Constr Build Mater 113:1029–1033
11. García A, Schlangen E, van de Ven M, Liu Q (2012) A simple model to define induction heating in asphalt mastic. Constr Build Mater 31:38–46
12. Li B (2018) Induction heating and healing characteristics of asphalt mixture. Master thesis. Wuhan University of Technology, China
13. Liu Q, Li B, Schlangen E, Sun Y, Wu S (2017) Research on the mechanical, thermal, induction heating and healing properties of steel slag/steel fibers composite asphalt mixture. Appl Sci-Basel 7:1088
14. Liu Q, Yu W, Wu SP, Schlangen E, Pan P (2017) A comparative study of the induction healing behaviors of hot and warm mix asphalt. Constr Build Mater 144:663–670
15. Liu Q, Chen C, Li B, Sun Y, Li H (2018) Heating characteristics and induced healing efficiencies of asphalt mixture via induction and microwave heating. Materials 11(6):913
16. Liu Q, Schlangen E, van de Ven M (2012) Evaluation of the induction healing effect of porous asphalt concrete through four-point bending fatigue test. Constr Build Mater 29:403–409
17. NL Agency Ministry of Economic Affairs, Agriculture and Innovation of the Netherlands (2011) Self healing Materials: concept and applications, 2nd ed
18. Partl MN (2018) Introduction. RILEM state-of-the-art reports. In: Partl MN, Porot L, Di Benedetto H, Canestrari F, Marsac P, Tebaldi G (eds) Testing and characterization of sustainable innovative bituminous materials and systems, vol 24, pp 1–14. Springer, Cham

19. Agzenai Y, Pozuelo J, Sanz J, Perez I, Baselga J (2015) Advanced self-healing asphalt composites in the pavement performance field: mechanisms at the nano level and new repairing methodologies. Recent Pat Nanotechnol 9(1):43–50
20. Santagata E, Baglieri O, Tsantilis L, Chiappinelli G (2015) Fatigue and healing properties of nano-reinforced bituminous binders. Int J Fatigue 80:30–39
21. Shi H, Magaye R, Castranova V, Zhao J (2013) Titanium dioxide nanoparticles: a review of current toxicological data. Part Fibre Toxicol 10(15):1–33
22. Begani AZ, Begani RK (2018) Is exposure to titanium dioxide nanoparticle associated with occupational lung cancer among titanium dioxide production workers? An emerging issue. Ann Occup Environ Med 6(2):50–61
23. Chernova T, Murphy FA, Galavotti S, Sun X-M, Powley IR, Grosso S, Schinwald A, Zacarias-Cabeza J, Dudek KM, Dinsdale D, Quesne JL, Bennett J, Nakas A, Greaves P, Poland CA, Donaldson K, Bushell M, Willis AE, MacFarlane M (2017) Long-fiber carbon nanotubes replicate asbestos-induced mesothelioma with disruption of the tumor suppressor gene Cdkn2a (Ink4a/Arf). Curr Biol 27(21):3302–3314
24. Qiu J, van de Ven MFC, Wu S, Yu J, Molenaar AAA (2009) Investigating the self healing capability of bituminous binders. Road Mater Pavement 10(1):81–94
25. Tabatabaee N, Shafiee MH (2012) Effect of organoclay modified binders on fatigue performance. In: 7th RILEM international conference on cracking in pavements. RILEM Bookseries, vol 4, pp 869–878. Springer, Dordrecht
26. Fang C, Yu R, Liu S, Li Y (2013) Nanomaterials applied in asphalt modification: a review. J Mater Sci Technol 29(7):589–594
27. Jeoffroy E, Koulialias D, Yoon S, Partl M, Studart A (2016) Iron oxide nanoparticles for magnetically-triggered healing of bituminous materials. Constr Build Mater 112:497–505
28. Santagata E, Baglieri O, Tsantilis L, Dalmazzo D, Chiappinelli G (2016) Fatigue and healing properties of bituminous mastics reinforced with nano-sized additives. Mech Time-Depend Mater 20:367–387
29. Wang YY, Su JF, Schlangen E, Han NX, Han S, Li W (2016) Fabrication and characterization of self-healing microcapsules containing bituminous rejuvenator by a nano-inorganic/organic hybrid method. Constr Build Mater 121:471–482
30. Pérez I, Agzenai Y, Pozuelo J, Sanz J, Baselga J, García A, Pérez V (2016) Self-healing of asphalt mixes, containing conductive modified bitumen, using microwave heating. In: 6th Eurasphalt & Eurobitume Congress, Prague
31. Ganjei MA, Aflaki E (2019) Application of nano-silica and styrene-butadiene-styrene to improve asphalt mixture self healing. Int J Pavement Eng 20(1):89–99
32. Wang Z, Dai Q, Guo S, Wang R, Ye M, Yap YK (2017) Experimental investigation of physical properties and accelerated sunlight-healing performance of flake graphite and exfoliated graphite nanoplatelet modified asphalt materials. Constr Build Mater 134:412–423
33. Jeoffroy E, Bouville F, Bueno M, Studart A, Partl M (2018) Iron-based particles for the magnetically-triggered crack healing of bituminous materials. Constr Build Mater 164:775–782
34. Yoo DY, Kim S, Kim M, Kim D, Shin H (2018) Self-healing capability of asphalt concrete with carbon-based materials. J Mater Res Technol 427
35. Li C, Wua S, Chen Z, Tao G, Xiao Y (2018) Improved microwave heating and healing properties of bitumen by using nanometer microwave-absorbers. Constr Build Mater 189:757–767
36. Garcia A, Schlangen E, Ven M Van De (2010) Two ways of closing cracks on asphalt concrete pavements: microcapsules and induction heating. Key Eng Mater 573–576
37. Liu Q, García Á, Schlangen E, Van De VM (2011) Induction healing of asphalt mastic and porous asphalt concrete. Constr Build Mater 25:3746–3752
38. Menozzi A, Garcia A, Partl MN, Tebaldi G, Schuetz P (2015) Induction healing of fatigue damage in asphalt test samples. Constr Build Mater 74:162–168
39. Norambuena-Contreras J, Serpell R, Valdés G, Gonzalez A, Schlangen E (2016) Effect of fibres addition on the physical and mechanical properties of asphalt mixtures with crack-healing purposes by microwave radiation. Constr Build Mater 127:369–382

40. Norambuena-Contreras J, Garcia A (2016) Self-healing of asphalt mixture by microwave and induction heating. Mater Des 106:404–414
41. Von Hippel AR (1954) Dielectric materials and applications. Wiley, New York
42. Metaxas AC, Meredith RJ (1983) Industrial microwave heating. The Institution of Engineering and Technology, Herts, UK
43. Meredith RJ (1998) Engineer's handbook of industrial microwave heating. The Institution of Electrical Engineers, London, UK
44. Jaselskis EJ, Grigas J, Brilingas A (2003) Dielectric properties of asphalt pavement. J Mater Civ Eng 15:427–434
45. Gallego J, Del Val MA, Contreras V, Páez A (2013) Heating asphalt mixtures with microwaves to promote self-healing. Constr Build Mater 42:1–4
46. García A, Norambuena-Contreras J, Pacheco-Bueno M, Partl MN (2015) Single and multiple healing of porous and dense asphalt concrete. J Intell Mater Syst Struct 26:425–433
47. Norambuena-Contreras J, Gonzalez-Torre I (2017) Influence of the microwave heating time on the self-healing properties of asphalt mixtures. Appl Sci 7:1076
48. Franesqui MA, Yepes J, García-González C (2017) Top-down cracking self-healing of asphalt pavements with steel filler from industrial waste applying microwaves. Constr Build Mater 149:612–620
49. González A, Norambuena-Contreras J, Storey L, Schlangen E (2018) Effect of RAP and fibers addition on asphalt mixtures with self-healing properties gained by microwave radiation heating. Constr Build Mater 159:164–174
50. Abreu LPF, Oliveira JRM, Silva HMRD, Fonseca PV (2015) Recycled asphalt mixtures produced with high percentage of different waste materials. Constr Build Mater 84:230–238
51. Arabani M, Tahami SA (2017) Assessment of mechanical properties of rice husk ash modified asphalt mixture. Constr Build Mater 149:350–358
52. Poulikakos LD, Papadaskalopoulou C, Hofko B, Gschösser F, Cannone FA, Bueno M, Arraigada M, Sousa J, Ruiz R, Petit C, Loizidou M, Partl MN (2017) Harvesting the unexplored potential of European waste materials for road construction. Resour Conserv Recycl 116:32–44
53. Sun D, Sun G, Du Y, Zhu X, Lu T, Pang Q, Shi S, Dai Z (2017) Evaluation of optimized bio-asphalt containing high content waste cooking oil residues. Fuel 202:529–540
54. García A, Pacheco-Bueno M, Norambuena-Contreras J, Partl MN (2013) Induction healing of dense asphalt concrete. Constr Build Mater 49:1–7
55. González A, Valderrama J, Norambuena-Contreras J (2019) Microwave crack healing on conventional and modified asphalt mixtures with different additives: an experimental approach. Road Mater Pavement Des 1–14
56. Karimi MM, Jahanbakhsh H, Jahangiri B, Moghadas NF (2018) Induced heating-healing characterization of activated carbon modified asphalt concrete under microwave radiation. Constr Build Mater 178:254–271
57. Norambuena-Contreras J, Gonzalez A, Concha JL, Gonzalez-Torre I, Schlangen E (2018) Effect of metallic waste addition on the electrical, thermophysical and microwave crack-healing properties of asphalt mixtures. Constr Build Mater 187:1039–1050
58. González A, Norambuena-Contreras J, Storey L, Schlangen E (2018) Self-healing properties of recycled asphalt mixtures containing metal waste: an approach through microwave radiation heating. J Environ Manage 214
59. Zhu X, Ye F, Cai Y, Birgisson B, Lee K (2019) Self-healing properties of ferrite- filled open-graded friction course (OGFC) asphalt mixture after moisture damage. J Clean Prod 232:518–530
60. Sun Y, Wu S, Liu Q, Zeng W, Chen Z, Ye Q, Pan P (2017) Self-healing performance of asphalt mixtures through heating fibers or aggregate. Constr Build Mater 150:673–680
61. Zhang B, Li J, Sun J, Zhang S, Zhai H, Du Z (2002) Nanometer silicon carbide powder synthesis and its dielectric behavior in the GHz range. J Eur Ceram Soc 22:93–99
62. Garcia A, Jelfs J, Austin CJ (2015) Internal asphalt mixture rejuvenation using capsules. Constr Build Mater 101:309–316

63. Al-Mansoori T, Norambuena-Contreras J, Micaelo R, Garcia A (2018) Self-healing of asphalt mastic by the action of polymeric capsules containing rejuvenators. Constr Build Mater 161:330–339
64. Garcia A, Schlangen E, van de Ven M, Sierra-Beltran G (2010) Preparation of capsules containing rejuvenators for their use in asphalt concrete. J Hazard Mater 184(1–3):603–611
65. Garcia A, Schlangen E, van de Ven M (2010) Properties of capsules containing rejuvenators for their use in asphalt concrete. Fuel 90(2):583–591
66. Garcia A, Austin CJ, Jelfs J (2016) Mechanical properties of asphalt mixture containing sunflower oil capsules. J Clean Prod 118:124–132
67. Su J, Qiu J, Schlangen E (2013) Stability investigation of self-healing microcapsules containing rejuvenator for bitumen. Polym Degrad Stabil 98:1205–1215
68. Su JF, Qiu J, Schlangen E, Wang Y (2014) Experimental investigation of self-healing behaviour of bitumen/microcapsule composites by a modified beam on elastic foundation method. Mater Struct 1–10
69. Yang P, Han S, Su J, Wang YY, Zhang XL, Han NX, Li W (2017) Design of self-healing microcapsules containing bituminous rejuvenator with nano-$CaCO_3$/organic composite shell: mechanical properties, thermal stability, and compactability. Polym Composite 39
70. Tabakovic A, Post W, Garcia S, Schlangen E (2016) Compartmented alginate fibres as a healing agent (rejuvenator) delivery system and reinforcement for asphalt pavements. In: Eighth international conference on maintenance and rehabilitation of pavements
71. Aguirre MA, Hassan M, Shirzad S, Cooper Jr S, Negulescu I, Mohammad LN (2018) Evaluation of hollow-fibers encapsulating a rejuvenator in asphalt binder with recycled asphalt shingles. In: Transportation research board 97th annual meeting transportation research board
72. Tabaković A, Braak D, van Gerwen M, Copuroglu O, Post W, Garcia S, Schlangen S (2017) The compartmented alginate fibres optimisation for bitumen rejuvenator encapsulation. J Traffic Transp Eng 4:347–359
73. Norambuena-Contreras J, Yalçin E, Garcia A, Al-Mansoori T, Yilmaz M, Hudson-Griffiths R (2018) Effect of mixing and ageing on the mechanical and self-healing properties of asphalt mixtures containing polymeric capsules. Constr Build Mater 175:254–266
74. Al-Mansoori T, Micaelo R, Artamendi I, Norambuena-Contreras J, Garcia A (2017) Microcapsules for self-healing of asphalt mixture without compromising mechanical performance. Constr Build Mater 155:1091–1100
75. Norambuena-Contreras J, Yalcin E, Hudson-Griffiths R, Garcia A (2019) Mechanical and self-healing properties of Stone Mastic Asphalt containing encapsulated rejuvenators. J Mat Civil Eng 31(5):04019052
76. Al-Mansoori T, Norambuena-Contreras J, Garcia A (2018) Effect of capsule addition and healing temperature on the self-healing potential of asphalt mixtures. Mater Struct 51:53
77. Norambuena-Contreras J, Liu Q, Zhang L, Wu S, Yalcin E, Garcia A (2019) Influence of encapsulated sunflower oil on the mechanical and self-healing properties of dense-graded asphalt mixtures. Mater Struct 52:78
78. Tabakovic A, Schlangen E (2018) Self-healing asphalt for road pavements. In: 4th International conference on service life design for infrastructures (SLD4). Delft University of Technology, Delft, The Netherlands
79. Mercadé-Prieto R, Zhang Z (2012) Mechanical characterization of microspheres, capsules, cells and beads: a review. J Microencapsul 29(3):277–285
80. Walter J, Salsac AV, Barthes-Biesel D, Le Tallec P (2010) Coupling of finite element and boundary integral methods for a capsule in a Stokes flow. Int J Numer Meth Eng 83:829–850
81. Müller M, Schirm S, Teschner M, Heidelberger B, Gross M (2004) Interaction of fluids with deformable solids. Comp Animat Virt W 15(3–4):159–171
82. Alexiadis A (2015) The discrete multi-hybrid system for the simulation of solid-liquid flows. PLoS ONE 10(5):1–26
83. Ariane M, Wen W, Vigolo D, Brill A, Nash FGB, Barigou M, Alexiadis A (2017) Modelling and simulation of flow and agglomeration in deep veins valves using discrete multi physics. Comput Biol Med 89:96–103

84. Wu K, Yang D, Wright N (2016) A coupled SPH-DEM model for fluid-structure interaction problems with free-surface flow and structural failure. Comput Struct 177:141–161
85. Ariane M, Vigolo D, Brill A, Nash FGB, Barigou M, Alexiadis A (2018) Using discrete multi-physics for studying the dynamics of emboli in flexible venous valves. Comput Fluids 166:57–63
86. Van Liedekerke P, Tijskens E, Ramon H, Ghysels P, Samaey G, Roose D (2010) Particle-based model to simulate the micromechanics of biological cells. Phy Rev E Stat, Nonlin Soft Matter Phys. 81(6):1–15
87. Wacker B, Kalantari M, Beatens B, Bochove van G (2018) HEALROAD - D5.2: induction heating asphalt mixes to increase road durability and reduce maintenance costs and disruptions - Evaluation of accelerated pavement test and analysis of structural changes
88. Gomez-Meijide B, Bianca Baetens B, van Bochove GG, Castro D, Garcia A, Indacoechea I, Lastra-Gonzalez P, Lizasoain E, Bastian Wacker B (2018) HEALROAD Project: laboratory optimisation and on-site validation of induction healing asphalt. In: 72nd RILEM annual week and 4th international conference on service life design for infrastructures (SLD4), Delft, The Netherlands
89. Steyn WJ (2012) NCHRP Synthesis 433: significant findings from full-scale accelerated pavement testing, Washington D.C.: Transport Research Board
90. Xu S, García A, Su J, Liu Q, Tabaković A, Schlangen E (2018) Self-healing asphalt review: from idea to practice. Adv Mater Interfaces 5(17), art. no.1800536

Multiscale Measurements of the Self-Healing Capability on Bituminous Materials

Guoqiang Sun, Daquan Sun, Alvaro Guarin, and Jose Norambuena-Contreras

Abstract This chapter aims to present a relatively comprehensive review for the self-healing measurements of bituminous materials based on the multiscale view. The application of the multiscale characterization technology can not only provide the diverse results of influencing factors on the self-healing capability of bituminous materials, but also to supplement in-depth insights into comprehending the self-healing mechanism from various scales. The multiscale analysis to the complex physical and chemical interactions between the base bitumen and other inclusions (such as fillers, aggregates, polymers, and so on) can help to reveal the self-healing mechanisms.

1 Introduction

Fatigue crack and even crack failure of bituminous mixture caused by repeated driving loads and environmental coupling aging (oxidation, light, water, etc.) are

G. Sun
Beijing Key Laboratory of Traffic Engineering, Beijing University of Technology, Beijing, P.R. China
e-mail: gqsun@bjut.edu.cn

D. Sun (✉)
Key Laboratory of Road and Traffic Engineering of Ministry of Education, Tongji University, Shanghai, P.R. China
e-mail: dqsun@tongji.edu.cn

A. Guarin
Department of Civil and Architectural Engineering, KTH Royal Institute of Technology, Stockholm, Sweden
e-mail: alvaro.guarin@abe.kth.se

J. Norambuena-Contreras
LabMAT, Department of Civil and Environmental Engineering, University of Bío-Bío, Concepción, Chile
e-mail: jnorambuena@ubiobio.cls

© Springer Nature Switzerland AG 2022
A. Kanellopoulos and J. Norambuena-Contreras (eds.), *Self-Healing Construction Materials*, Engineering Materials and Processes,
https://doi.org/10.1007/978-3-030-86880-2_5

the main damage forms of bituminous pavement. The fatigue damage of bituminous mixture is a complex process of crack initiation, development, aggregation and finally the formation of macro cracks. The cracking behavior of bituminous materials has multiscale feature in nature [1]. The nano/micro-scale (approximately 10^{-9} m) cracking is presented as crystal defect or molecular aggregate rupture. The meso-scale cracking (approximately 10^{-5} m) appears as micro-cracks or micro-holes in continuous body, or interface cracks in composite structures. As regard to the macro-scale ($>10^{-3}$ m) cracking, it can be detected by human eyes. Figure 1 presents various cracking types in bituminous materials. The cracking at pavement level includes bottom-up cracks as a result of tensile strains or top-down cracks caused by tensile and shear stresses coupling with environmental effects [2]. The cracks in bituminous mixture specimens are formed by repeated fatigue loading. Further, the cracks in bitumen/mastics/interface can be classified as cohesive cracks in bitumen/mastics or adhesive cracks at bitumen–aggregate interface.

By contrast, self-healing of bituminous materials is the reverse process of fatigue damage, that is, cracks are closed spontaneously and strength is restored gradually. There are two occurring ways for macro-scale self-healing behavior. One way is that the pavement cracks are healed in certain rest time. Another way is that pavement cracks created in winter are healed in warm summer. The meso-scale self-healing behavior is manifested as cohesive self-healing in bitumen or mastics or adhesive self-healing occurring at interface. The micro-scale self-healing process mainly includes the spontaneous approaching, contacting, wetting and diffusing of bitumen molecules [3–5].

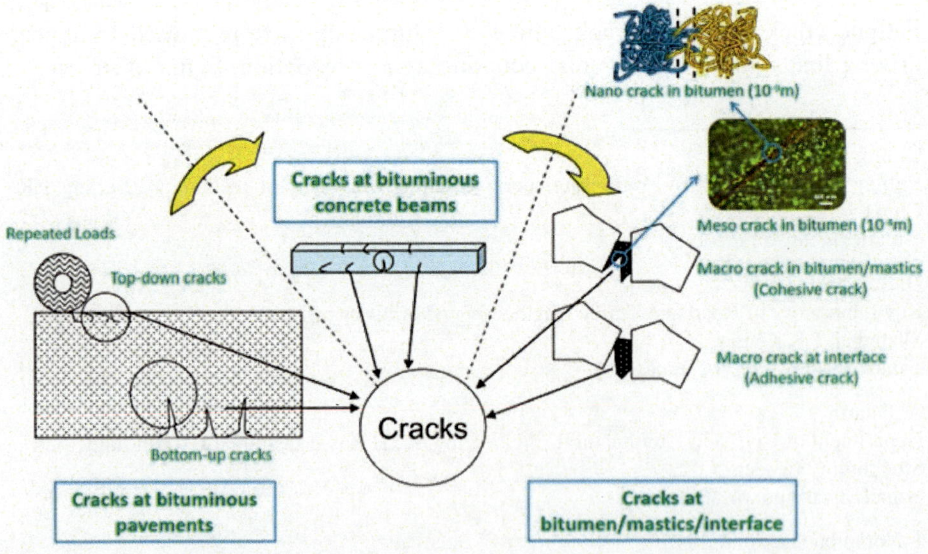

Fig. 1 Various cracks/self-healing types of bituminous materials based on multiscale view (Modified from Qiu [2])

To monitor and assess the self-healing capability (including the self-healing index or self-healing rate based on mechanical or other physicochemical recovery) in bitumen binder, mastics, mortars/mixtures and pavement, a mass of methods based on various scales have been proposed. The primary ways assessing self-healing ability is to identify the response differences by designing damage conditions with varying rest periods and temperatures [2, 6]. However, there is no unified evaluation method and standard to quantify the self-healing ability of bituminous materials. Even some conclusions on healing performance for the same bituminous material are contradictory. Furthermore, the differences of self-healing capability are also significant between the binder and its mixture since some examined healing at bitumen level, whereas others assessed healing at mixture or mastics level [7]. The following sections aim to present a relatively comprehensive review for the self-healing measurements of bituminous materials based on the multiscale view.

2 Micro-scale Self-Healing Measurements

The micro-scale self-healing measurements of bituminous materials mainly rely on scanning electron microscopy (SEM) and atomic force microscopy (AFM). They are applied to examine the micro-scale self-healing process by producing artificial cracks in bituminous materials [8]. The modified SEM technique is introduced to look into the bitumen surface morphology since the bitumen samples are observed with a metal coating film for conventional SEM.

Lu and co-workers [9] assessed the influence of cracking width on self-healing via the field emission SEM (FM-SEM). Shen et al. [10] further used FM-SEM to investigate the aging impact on the crack self-healing process of bitumen, as shown in Fig. 2. By contrast, AFM can not only display more realistic and higher resolution surface morphology, but also be used to detect the mechanical performances of materials from nano-scale to micro-scale [11–14]. Loeber et al. firstly obtained the asphalt AFM micrograph with various microstructures [15]. Base on AFM observation, bitumen can be divided as two phases, i.e., stiffer domain phase (or called as bee-structure) and softer matrix phase. In fact, bitumen is a single-phase melt-state liquid over 80 °C, while phase separates once cooling. With the progress of phase separation, viscosity in bitumen begins to accumulate, leading to the bitumen state changing from liquid to semi-solid or solid. The heterogeneity of the bitumen microstructure property was also found via differential scanning calorimetry (DSC) and thermo-microscopy [16]. Moreover, Nahar [17] attempts to build the relationship between phase separation and damage healing of asphalt. It was observed that micro-cracks initiated at the phase interfaces and propagated inward the domain phase at moderate loading condition; however, under high-loading condition, micro-cracks coalesced and led to fragmented domain phases. Further, the cracks in microstructure were gradually restored after moderate heating, as shown in Fig. 3.

Combined with phase field theory and micro-finite element method, AFM becomes a favorable selection to investigate the microscopic morphology and

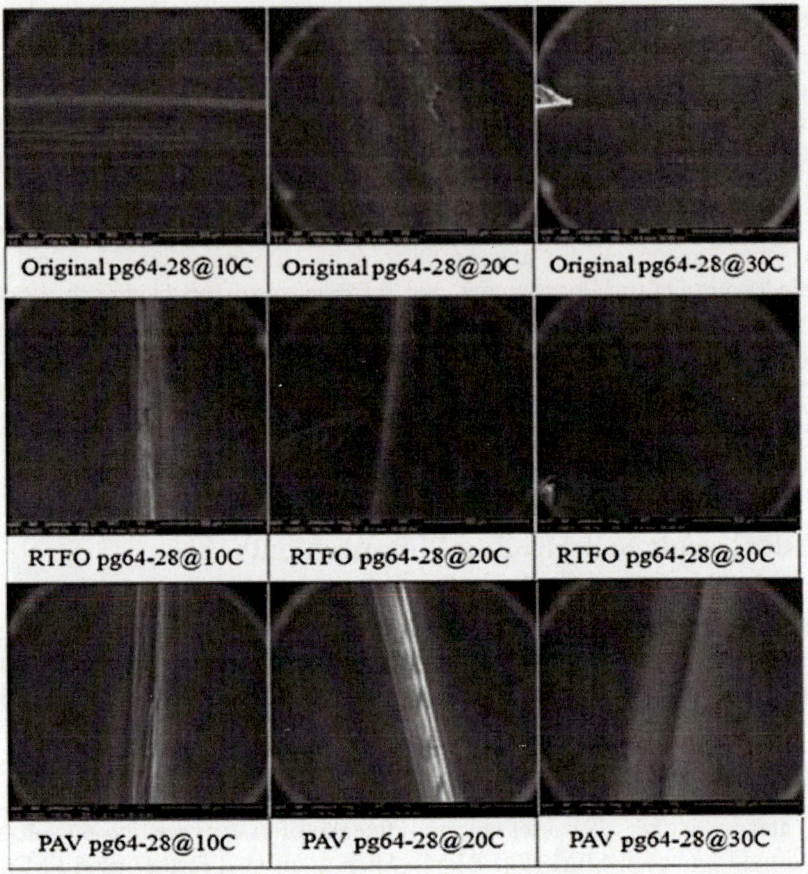

Fig. 2 Self-healing of aged binder via FM-SEM with 3 h' rest at 10 °C, 20 °C, and 30 °C (Reproduced from Shen et al. [10])

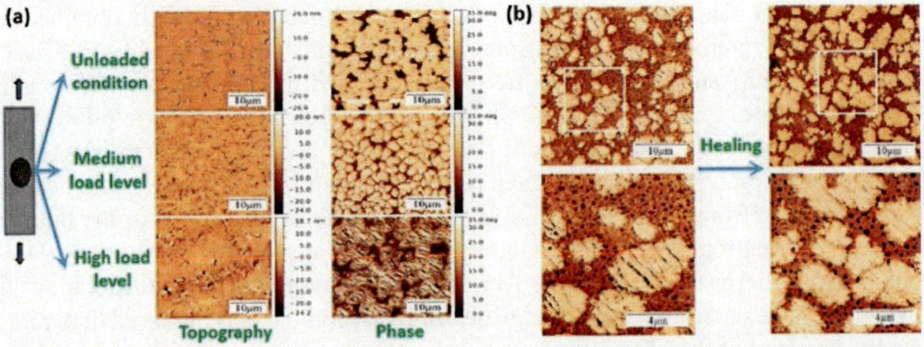

Fig. 3 AFM micrographs of bitumen microstructures: **a** AFM-topography and AFM-phase under various load conditions; **b** micro-cracks self-healing in microstructures (Modified from Nahar [17])

damage self-healing characteristics of bitumen. The heterogeneity and physiochemical property of bitumen microstructures determine the cracking behavior of bitumen; however, the distribution of bulk microstructure in asphalt and the relationships among the chemical components, inter-molecular interaction and microstructures deserve insight research. Therefore, it is expected that another advanced micro-scale observation, neutron scattering, can be used to examine the mobility of the microstructures, so as to comprehend the self-healing mechanism of bitumen.

Moreover, some indirect observations by using microscopic characterization methods, such as Fourier transform infrared spectroscopy (FTIR), gel permeation chromatography (GPC), thin-layer chromatography (TLC), nuclear magnetic resonance (NMR), and flame ionization detection (FID), have also played an essential role in the self-healing research of bituminous materials. Numerous studies reflect that the intrinsic self-healing ability of bitumen is closely related to its chemical properties. Kim et al. [18] found that two molecular indices measured by FTIR and NMR, i.e., the methyl plus methylene hydrogen to carbon (MMHC) ratio and the CH_2/CH_3 (methylene to methyl) ratio, can characterize the branched-chain number and their length, respectively. Smaller MMHC ratio and greater CH_2/CH_3 ratio mean the longer and thinner average molecular structure of bitumen (with less branched-chains and higher molecular mobility). In this way, the bitumen molecules with greater mobility will diffuse between the crack faces more smoothly, thus promoting the self-healing rate. The investigation conducted by Williams et al. [19] presented that the bitumen with low amphoteric and high aromatic contents represents greater self-healing ability. Santagata et al. [20] discovered that the pure bitumen with more low molecular weight (aromatics) components shows higher self-healing ability using TLC and FID. Further, Sun et al. [21] analyzed the influence of chemical property on the self-healing performance of bitumen using TLC, GPC, FTIR, NMR and fatigue-healing-fatigue tests. It was confirmed that the bitumen with higher small molecule content/large molecule content ratio or higher aromatics content has stronger self-healing ability, and the bitumen molecule with less branched-chains and longer structures is conductive to self-healing.

3 Meso-scale Self-Healing Measurements

It has been widely confirmed that self-healing behavior essentially includes the approach, wetting, diffusion and randomization of molecules at the crack interface [2, 4]. However, the self-healing process is manifested as bitumen flow and crack closure on the meso-scale view, and its external impacts can be examined via mesoscopic means, such as the fluorescence microscopy (FM) and X-ray computed tomography (X-ray CT). For example, Sun et al. [22] identified the crack self-healing behavior of SBS modified asphalt through FM, and proposed the self-healing index according to the variation of meso-crack area with the software Image-Pro Plus 6.0. Two critical self-healing stages of bitumens, i.e., wetting and diffusion, were differentiated by the proposed two-stage model. Figure 4 presented the captured self-healing images of

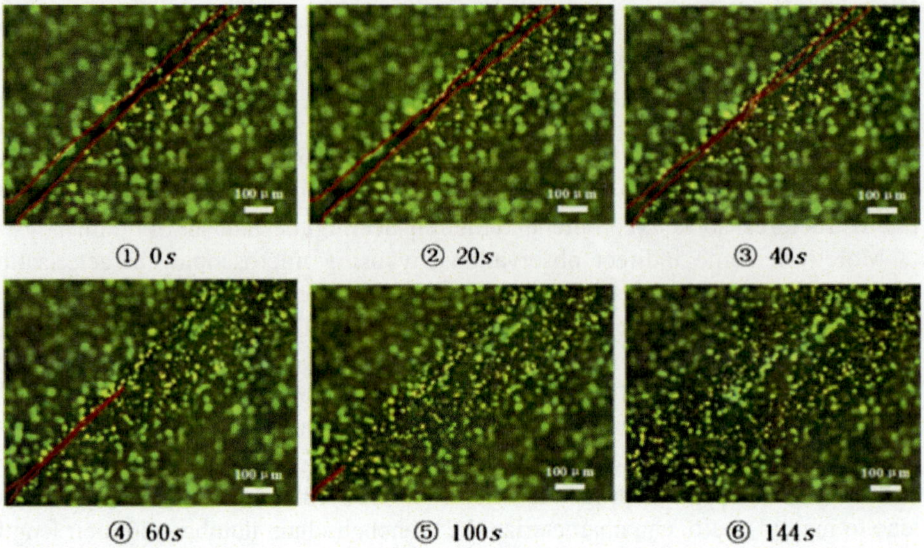

Fig. 4 Crack self-healing process of SBS modified bitumen via FM observation (Reproduced from Sun et al. [22])

styrene—butadiene-styrene (SBS) modified bitumen. In addition, the further study on the use of X-ray CT to assess the self-healing of bituminous mastics samples presented 2D and 3D images of the internal structure of objects at various damage stages [23–25]. Menozzi et al. [26] studied the variation of void ratio in bituminous mixture samples before and after induction heating for 10 s through CT scanning test. García [27] used CT scanning technology to observe the induction heating self-healing process every 5 or 10 min based on the three-point bending fracture-healing test, see Fig. 5.

Notably, the improved X-ray CT or FM apparatuses are expected to be equipped with real-time load apparatus and thermostatic system, so as to obtain the timely crack self-healing information at various test conditions (such as loading types and temperature). Although the mesoscopic measurements exhibits a clear path to examine the crack self-healing behavior of bituminous material specimens at various conditions, the in-depth analysis on the test data is worth further research together with the digital image processing technique, statistical methods and numerical simulation technology. For example, the numerical simulation model on the basis of CT reconstruction of bituminous mixture could be used to accurately simulate the cracking and self-healing behaviors, so as to predict the mechanical performance and fatigue span.

In this context, Norambuena-Contreras and Garcia [28] used X-ray computed tomography to monitor variations of internal structures (cracking damages) in the bituminous mixture samples before, and after, microwave heating. The X-ray micro-tomography scanning was conducted via the XRadia Versa XRM-500 scanner (80 kV, 120 μA). According to the simple threshold method, the specific volume of material

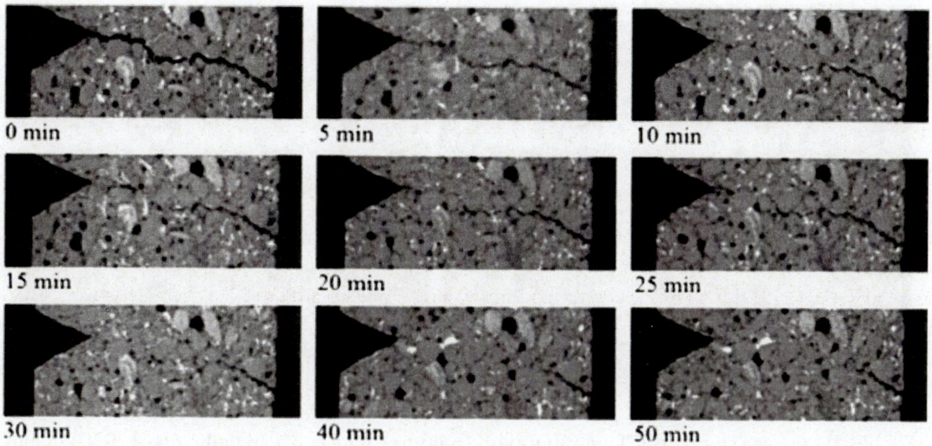

Fig. 5 Crack self-healing process of bituminous mastics via CT scanning technique (Reproduced from García [27])

was segmented, so as to prepare the air void reconstruction model using the reconstruction software ImageJ® [28]. Aggregate, fiber, bitumen, and voids were further separated by this method. Examples of CT scanning images of the sample before and after microwave heating are shown in Fig. 6, which shows the air void distribution in the test specimen: due to microwave heating, voids are redistributed in the specimen, increasing in some areas and decreasing in other areas.

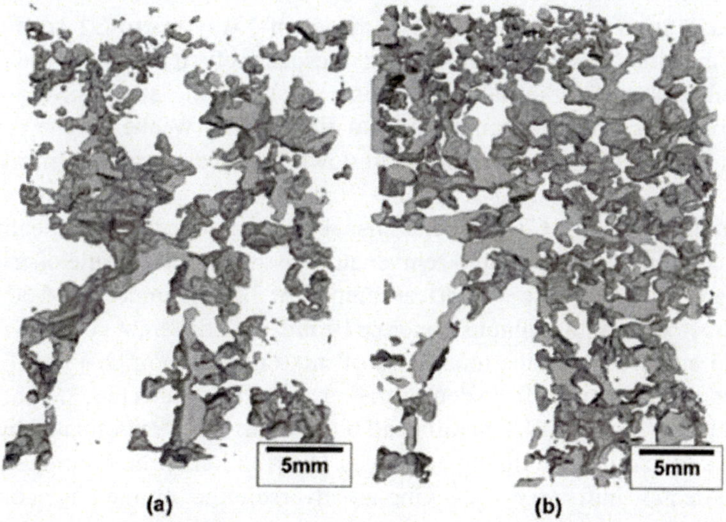

Fig. 6 A3D CT scan images of the air voids distribution into dense-graded asphalt mixture samples containing 4% of steel wool fibers: **a** before microwave radiation heating, and **b** after microwave radiation heating (Modified from Norambuena-Contreras and García [28])

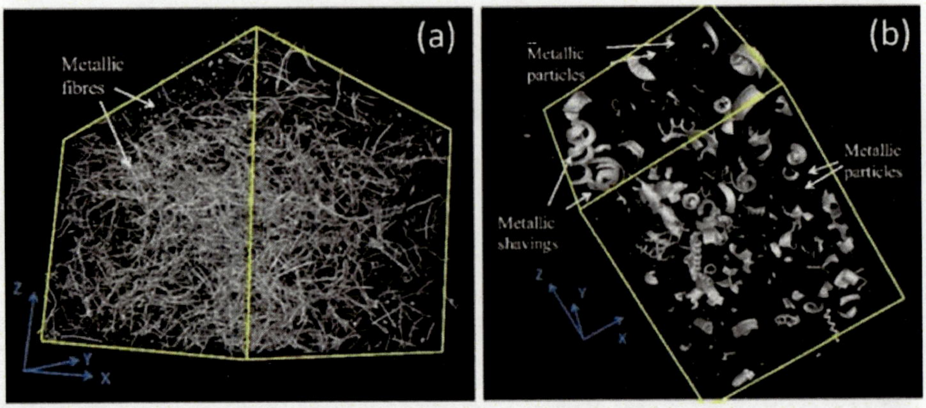

Fig. 7 3D images of micro-CT on bituminous mixtures with **a** 4% metallic fibers, **b** 4% metallic shavings (Reproduced from Norambuena-Contreras et al. [33])

Furthermore, several studies published by Norambuena-Contreras et al. have used CT scanning to study the spatial distribution of the metallic fibers in the bituminous mixture with self-healing goals by microwave induction [29, 30]. And also to evaluate the meso and microstructure of sustainable asphalt mixtures containing various metal waste [31–33]. In this context, in work [33] X-ray micro CT scanning was carried out on bituminous mixture samples containing four various contents of metallic waste: 2%, 4%, 6%, and 8% by the total bitumen volume into the mixture. To do the CT scans, prismatic samples with dimensions 20 × 20 × 50 mm were cut from the center of semi-circular specimens with 4% metallic waste after the crack self-healing. Figure 7a, b shows an example of 3D images of X-ray micro-CT scan of samples with 4% metallic fibers and steel shavings, respectively. It was observed from Fig. 6 that the fibers keep their initial morphology, whereas some steel shavings broke into smaller particles. The mechanical degradation overall weakened the capability of shavings to form clusters, and hence cut down the voids formation inside mixtures [33].

Additionally, X-ray CT scan has also been employed in self-healing asphalt mixture containing encapsulated rejuvenators to assess the capsule distribution and their integrity [34, 35]. The cylindrical samples of 3.5 cm diameter and 60 mm height were extracted from bituminous concrete beams with different contents of capsules (0.5, 0.75 and 1.00% by the total mass of mixture). In short, X-ray CT scan were employed on test samples from dense-graded asphalt mixtures [34, 35] and also stone mastic asphalt (SMA) [36]. Overall in all bituminous mixture evaluated by X-ray CT tests allowed to conclude that the capsules' distribution in the bituminous mixtures was regarded as uniform, not showing an adverse effect on the physicomechanical performances studied.

4 Macro-scale Self-Healing Measurements

The macro-scale self-healing measurements are mainly fatigue-based self-healing tests, fracture-based self-healing tests, and field-based self-healing tests. Most macro-scale measurement means are mechanical based experiments. Table 1 lists the main test methods for the macro-scale self-healing studies of bituminous materials.

4.1 *Fatigue-Based Self-Healing Tests*

Table 2 presents the main fatigue-based self-healing experiments on the bituminous materials including bitumen, mastics, mortars, and mixtures.

From Table 2, it can be concluded that the fatigue-based self-healing tests are carried out at two kinds of load types, i.e., intermittent type and interruption type [2, 6]. A detailed description of these tests is discussed below.

Table 1 Macro-scale self-healing measurements of bituminous materials

Test methods		Bitumen/mastics/mortars level	Bituminous mixtures level	Bituminous pavement level
Fatigue-based self-healing tests	Intermittent test	Dynamic shear rheometer (DSR), dynamic mechanical analysis (DMA)	Uniaxial tensile test (UT), three-point bending test (3 PB), four-point bending test (4 PB), semi-circular bending test (SCB), Superpave indirect tensile test (IDT)	—
	Interruption test			
Fracture-based self-healing tests		DSR, binder bond strength test (BBS)	3 PB, SCB	—
Field-based self-healing tests	—			Falling weight deflectometer (FWD); Surface wave (SW) test;

4.1.1 The Intermittent Tests

There is a rest period after each intermittent loading cycle. Intermittent loading tests seems to be closer to the real scene of road self-healing since there is a gap between the loading of continuous axles of passing vehicles. This mode was usually used in SCB,

Table 2 Fatigue-based self-healing tests on bituminous materials

Authors	Material	Self-healing temperature (°C)	Load modes	Self-healing process
Lu et al. [37]	Bitumen	15	Stress control mode (DSR); 184–237 kPa	Intermittent: 20 s load along with10s rest and 20 s load along with 400 s rest
Shen et al. [38]	Bitumen	15, 25	Stress control mode (DSR); 60–230 kPa	Intermittent: 1 s load along with 0 s rest and 1 s load along with 6 s rest
Van den bergh et al. [39]	Mastics/Mortars	15	Stress control mode (DSR); 0.025–0.1 Nm	Intermittent: 3 s load along with 9 s rest
Sun et al. [40]	Mixture	−5, 5	Stress control mode (IDT); 0.30 St at 5 °C, and 0.50 St at −5 °C	Intermittent: 0.1 s load along with 1 s rest & 0.1 s load along with 3 s rest
Castro & Sánchez [41]	Mixture	20	Strain control mode (3 PB); 200–400 micro-strains	Intermittent: 0.1 s sinusoidal load with 1 s rest each cycle
Carpenter & Shen [42]	Mixture	20	Strain control mode (4 PB); 500 micro-strains	Intermittent: 0.1 s haversine load with 0 s, 3 s, 9 s rest each cycle
Jiang et al. [43]	Mixture	15	Stress control mode (SCB); 5kN	Intermittent: 0.1 s haversine load with 0 s, 0.4 s, 0.8 s, and 1.2 s rest each cycle
Xie et al. [44]	Bitumen	10, 20, 30	Strain amplitude ramping from 0.1% to 30% in a step manner	Interrupted: the maximum stored pseudo-strain energy, rest for 1 min, 5 min, 15 min, and 30 min

(continued)

Table 2 (continued)

Authors	Material	Self-healing temperature (°C)	Load modes	Self-healing process
Li et al. [45]	Bitumen	20	Strain control mode (DSR); 5%, 7%, and 10%	Interrupted: after 5 min, 10 min and 20 min, rest for 5 s, 10 s, 30 s, 1 min, 2 min, 5 min, 10 min, 20 min, and 40 min
Bahia et al. [46]	Bitumen	20	Strain control mode (DSR); 20%	Interrupted: 5000 loading cycles, rest for 0.5 h, 1 h, 3 h, and 12 h
Sun et al. [40]	Bitumen	15–50 every 5 °C	Strain control mode (DSR); 2–6% dependent on temperature	Interrupted: after 70% modulus reduction, rest for 1 min, 5 min, 30 min, and 60 min
Sun et al. [40]	Mastics	15, 20, 23	Stress control mode (DSR); 0.5 MPa	Interrupted: rest for 1 h
Smith & Hesp [47]	Mastics	10	Strain control mode (DSR); 0.3%	Interrupted: rest for 2 h
Li et al. [48]	Mortars	25	Stress control mode (DMA); constant amplitude oscillatory shear stress at 5 Hz	Interrupted: 80% modulus reduction, rest for 5 min, 10 min, 20 min, and 40 min
Kim et al. [49]	Mixture	10–20	Stress control mode (IDT); 700–1300 lb, 55, 75psi	Interrupted: 0.1 s load with 0.9 s rest each loading cycle, conduct modulus tests at 2 min, 4 min, 6 min, 10 min, 20 min, 40 min, and 60 min after 1000 cycles' loads
Baaj et al. [50]	Mixture	20	Strain control mode (UT fatigue test); 80–100 μm/m	Interrupted: load 4-8 h, rest for 16-20 h

(continued)

Table 2 (continued)

Authors	Material	Self-healing temperature (°C)	Load modes	Self-healing process
Abo-Qudais & Suleiman [51]	Mixture	0, 15, 30, 45 and 60	Stress control mode (IDT); 3.5kN	Interrupted: 0.1 s load with 0.4 s rest each cycle
Little et al. [52]	Mixture	25	Strain control mode (uniaxial cyclic fatigue test); 200, 350 microstrains	Interrupted: 0.1 s load with 0.9 s rest each cycle
Lee et al. [53]	Mixture	25	Stress control mode (UT); 267 N–400 N	Interrupted: 0.2 s haversine each cycle, then rest for 20, 40, 80, 320, 1280 s

IDT, 3 PB, 4 PB, and UT experiments to assess the self-healing capability of bituminous mixture [2, 26, 42, 44, 54–56], and the DSR based intermittent loading was applied in the bituminous binder. Specifically, Carpenter and Shen [42] performed the 4 PB based intermittent test (10 Hz, 20 °C, rest periods (RPs): 0 s, 1 s, 3 s and 9 s) of bituminous mixtures at 500 microstrains. A new self-healing indicator, i.e., the plateau value (PV) based on the ratio of dissipated energy change (RDEC), was proposed and independent of the loading mode and condition. Results indicated that the self-healing influence was more remarkable at lower strain condition or with longer rest period, and the polymer modification extended the fatigue span.

Further, Shen and Carpenter [38] conducted a DSR based intermittent test (10 Hz, 15 °C & 20 °C, RPs: 0 s & 6 s) to analyze the self-healing property of neat bitumen and polymer modified bitumen using the slope of the $PV - RP + 1$ curve under log–log plot as the self-healing rate. Results also revealed that the inclusion of 6 s-RP could result in seven times longer fatigue span for neat binder but 17 times for modified binder, meanwhile the self-healing was influenced by the initial strain level. Moreover, Lu et al. [37] also conducted the DSR based intermittent loading experiments with various load/rest ratios each cycle to study the self-healing performance of bitumen. Results reflected that the effect of self-healing time on fatigue span prolongation was closely related to the bitumen type to a great extent. A different conclusion from Jiang et al. [43] presented that the most self-healing recovery happened with the 0.8 s RP compared with 0.4 s and 1.2 s on the basis of the SCB based intermittent test. Besides, it was discovered that the modified bituminous mixture had both better fatigue and self-healing performances than neat mixtures, whereas the SMA mixture had similar fatigue property but far greater self-healing property than dense graded mixture with the same binder.

Notably, the measured performance recovery may not accurately reveal the real self-healing ability of bituminous materials, because these indicators can not subtract the time-varying hybrid effects (such as self-heating and plastic deformation) in the repeated cyclic load process. Kim, Little, and Lytton [57–59] put forward the dissipative pseudo-strain energy (DPSE, the dissipated energy each loading cycle caused by the damage in specimen) to analyze the fatigue crack damage, and the variation rate of DPSE can be used to evaluate the fatigue crack life. They used DMA to conduct strain controlled intermittent tests (RP: 2 min) to study the self-healing performance of fine bituminous mixture under the inclusion of RPs [57, 58]. The diameter and height of DMA cylindrical specimen were 12 mm and 50 mm, respectively. Results illustrated that the rest period included at the early stage of fatigue test can remarkably prolong the fatigue span, indicating the rapid self-healing of micro-damage was occurring in the rest period. In addition, Si et al. [60] characterized the self-healing ability of bituminous mixture with the repeated cycle UT fatigue experiment (the cylindrical test-piece with diameter of 100 mm and height of 150 mm was used for measurement). With the inclusion of RPs, the increase of pseudo-stiffness or the decrease of DPSE accumulation is regarded as the self-healing of bituminous mixture.

4.1.2 The Interruption Tests

During interrupted loading, the bituminous specimens will endure successive repeated loading for certain time, unless the rest time is introduced in which the test-piece will remain under the given conditions without loading. Interruption intervals with various/constant rest time were inserted under certain damage degrees to observe the mechanical performance recovery (such as the modulus and energy). Many researchers have achieved this load mode by DSR, DMA, 3 PB, 4 PB, SCB, and IDT experiments to monitor the self-healing ability of bituminous materials [2, 6, 61].

Bahia et al. [46] studied the self-healing performance of modified bitumens via the interruption mode test at 20 °C and 1.6 Hz. The first phase was the fatigue test for 5000 cycles with the constant strain of 20%, then the RPs were set as 0.5 h, 1 h, 3 h, and 12 h. Each fatigue test would be restarted until the set damage level. Results revealed that the RP inclusion prolonged the total number of load cycles to failure. Santaga et al. [20] investigated that the relationship between the self-healing ability of neat bitumens and its chemical composition using the interruption test at 1.59 Hz and 1% strain. In the self-healing duration of 2 h, the specimens were subjected to 0.01% strain oscillation to examine the modulus recovery level. Results showed that the self-healing ability of bitumen with more light components was stronger. Recently, Sun et al. [62] analyzed the self-healing evolution behavior within a wide temperature range from 15 °C to 50 °C using the interruption mode-based fatigue—healing–fatigue (FHF) test. The initial RPs were set as 1 min and 5 min, and the longer RPs were set dependent on the temperature (60 min and 90 min at 15 °C ~ 25 °C, 30 min and 60 min at 30 °C ~ 35 °C, 15 min and 30 min at 40 °C ~ 50 °C). The proposed self-healing transition phenomenon revealed that the self-healing ability could not be characterized comprehensively and accurately at only single temperature. Moreover, Pronk and Cocrullo [63] adopted the 4 PB based interruption test under strain control mode at 5 Hz and 20 °C, and the load and RP were set as 40,000/400,000 cycles. Results showed that the almost complete modulus recovery can not ensure the extension of fatigue span since the intrinsic self-healing could not totally repair the fatigue crack damage in the specimen.

In addition, to simulate the loading condition of bituminous pavement as truly as possible, researchers developed the method of combining intermittent and intermittent loading to study the self-healing performance of bituminous materials. Abo-Qudais and Suleiman [51] performed the IDT fatigue-healing experiment (cyclic load of 3.5 kN, 0.1 s load/0.4 s rest each cycle) together with ultrasonic pulse velocity (UPV) test to analyze the self-healing and fatigue performances of bituminous mixture. The fatigue damage level was set at which the stiffness decreased to 3/4 of the initial value. After that, the samples were conditioned at five various temperatures (0, 15, 30, 45, and 60 °C) for various interval time (1 h, 3 h, 3, 7, and 14 days). The prolonged fatigue span was determined at the end of self-healing time by IDT fatigue test, and the UPV test was also applied. Results showed that UPV reduced with the increase of repeated load times and rise with the increase of self-healing time/temperature. Recently, Sun et al. [54] also conducted a IDT fatigue-healing test at lower temperatures (5 °C and −5 °C). The intermittent time (0 s, 1 s, and 3 s) and interruption time (2 min, 10 min, 30 min, and 60 min) were applied. Results presented that the longer the intermittent time, the less the impact on fatigue-healing of bituminous mixture, and the low temperature (−5 °C) remarkably decreased the self-healing rate of mixtures, and led to very low level long-term self-healing rate.

Moreover, the load waveforms (such as haversine load or sinusoidal load) varied in various fatigue-healing investigations. Grant [48], Kim and Roque [49] carried out the IDT fatigue-healing experiment at 10 °C, 15 °C, and 20 °C. Firstly, the discontinuous haversine loading (0.1 s load/0.9 s rest) was employed to the specimen for 1000 cycles. Then, the resilient modulus test was then performed at various rest time to determine the overall self-healing rate by calculating the damage recovery rate based on the proposed dissipative creep strain energy (DCSE) indicator. By contrast, Sun et al. [54] conducted the continuous IDT fatigue test (0.1 s haversine loading per cycle) to certain damage degree (i.e., reduction of stiffness), and the discontinuous haversine loadings (0.1 s load/1 s or 3 s rest) were applied for the same cycles with the fatigue test to determine the effect of intermittent time on the stiffness recovery. Besides, Mamlouk et al. [64] conducted the 4 PB fatigue test (strain controlled at 400 and 800 microstrains, 10 Hz) using the haversine load and sinusoidal load at 4 °C, 21 °C, and 38 °C. The intermittent modes with 5 s and 10 s RPs were used. Results reflected that the sinusoidal load could be regarded as a more consistent and precise way to investigate the fatigue and self-healing, while the haversine signal (stress or strain control) was hard to maintaine because of the viscoelasticity of bituminous materials.

4.2 Fracture-Based Self-Healing Tests

Fracture-based self-healing tests were performed to examine the self-healing responses of the recontact specimens through introducing rest time at certain conditioning temperature. Bazin and Saunier [65] firstly proposed the self-healing phenomenon in bituminous mixture by the fracture-based self-healing tests in 1967. They assessed the fracture self-healing behavior through the tensile tests. Bituminous mixture sample (4 cm × 3 cm × 10 cm) was fractured at 120 mm/min, and then, the two parts were contacted again, keeping perpendicular at 10 °C, 18 °C, and 25 °C for one day to 300 days. Results exhibited that the mixture type and temperature had a significant influence on the fracture self-healing process (based on the recovered fracture strength). It was also suggested that the permanent compressive stress could enhance self-healing ability. In recent studies, the SCB or 3 PB based fracture-healing tests were widely applied to assess the enhanced self-healing performances of bituminous mixtures through induction heating or encapsuled technology [66–68]. Jahanbakhsh et al. [66] performed the SCB based cracking-healing scenario. The specimen was firstly fractured at a monotonic load (0.6 mm/min, −20 °C) and was kept in contact again (25 °C for 24 h), and then, the microwave radiation was applied for 120 s. Lastly, the healed specimen was fractured again. Results confirmed the effects of temperature and rheological performances on the induced self-healing behavior in bituminous concrete. Al-Mansoori et al. [67] conducted the 3 PB-based crack healing test to examine the self-healing performance of bituminous mixture with capsules. Similarly, the mixture beam was broken at −20 °C (2 mm/min). Then, the broken parts were put together again into a steel mold (20 °C for 2 h), and later, a compressive loading was uniformly applied on the beam surface to accelerate the capsule release. Lastly, the healed specimen was conditioning for 0.3 ~ 8 days. Results reflected that the greater capsule content resulted in higher self-healing level, and the self-healing temperature had more significant self-healing effect below 40 °C.

At the bitumen/mastics level, Bhasin et al. [69] proposed the two-piece healing (TPH) experiment to directly evaluate the crack self-healing behavior of bitumen at 20 °C and 25 °C. For this purpose, two bitumen pieces were put on the upper plate and bottom plate of specimen holder in DSR. Then, the two bitumen pieces were spliced together with a small constant compression normal force of 0.4 N. This new method could be used to determine the parameters of intrinsic self-healing function. In addition, Qiu et al. [70] proposed a direct tensile fracture-healing test of mastics. The fracture test was performed at 100 mm/min, and the rest time was 3 h, 6 h, and 24 h, and the self-healing temperature was 10 °C, 20 °C, and 40 °C, respectively. Results show that the re-fractured strength increased with self-healing time and temperature. Recently, the BBS-based fracture-healing tests were introduced to characterize the adhesive self-healing property of bitumen binders or mastics [71, 72]. Lv et al. [71] applied a multi-failure-healing cycle test based on the BBS test to analyze the influences of aging and temperature on the bond strength recovery of bitumens. Results revealed that self-healing temperature was more significant than self-healing time, and aging led to various effects on the base binder and modified binder. Further, Zhou et al. [72] investigated the adhesive self-healing property of various modified bitumens via the BBS-based self-healing test. It was found that high-density polyethylene and rubber modifications could enhance the bitumen self-healing property.

In fact, the fracture-healing experiments reflected that visible cracks inside bituminous materials can be gradually closed, thus restoring stiffness and strength. There are complex inter-molecular motions like wetting and diffusion occurring between the cracking faces. Nevertheless, there were some controversies for the accuracy of evaluating the self-healing ability when employing the fracture-based self-healing tests in bituminous materials. One was the plastic deformation influence caused by the contact force on crack faces inevitably. Another problem was the uneven contact condition, because the fracture faces can not guarantee the complete contacts with each other because of errors and permanent deformation.

4.3 Field-Based Self-Healing Tests

Compared with laboratory studies, the field self-healing tests were limited. Two main measurement means were used to monitor the self-healing ability of bituminous pavement [73, 74], i.e., FWD test and SW test. Groenendijk [74] used the FWD experiment during accelerated pavement test. The successive repeated loading was stopped (4 million times of 75 kN load) for rest from the afternoon of every Friday to the morning of next Monday. It was shown that the stiffness recovery of pavement was produced because of the self-healing behavior in bituminous mixture. Williams et al. [19] studied the influence of rest time on pavement stiffness recovery through SW test. In this test, the SW transient generated by impact loading was used to examine the wave velocity in each layer which was correlated to the material characteristics. The SW and FWD tests provided the non-destructive testing techniques to analyze the self-healing capability of bituminous pavement. It was excepted that more accurate and faster non-destructive methods can be developed applied in the fields.

4.4 Self-Healing Index of Bituminous Materials

Early studies on self-healing capability of bituminous materials mainly focused on the bituminous mixtures. Bazin and Saunier [65] used the direct pull-up tests to study fatigue-healing behavior of bituminous mixture. After self-healing, the ratio of the tensile strength to the original tensile strength was used as an index to evaluate the self-healing ability.

In this context, Kim and co-workers [73] put forward the concept of self-healing index (SHI) as an indicator to quantify the self-healing capability of bituminous mixture. With this purpose, the bituminous materials were analyzed before and after self-healing test. Pseudo-dissipative strain energy ratio can be used as a judgment index for evaluating self-healing, see Eq. (1):

$$SHI = \frac{\Phi_{\text{after}} - \Phi_{\text{before}}}{\Phi_{\text{before}}} \tag{1}$$

where SHI was the self-healing index, and Φ_{before} and Φ_{after} represented the pseudo dissipative strain energy before and after healing, respectively.

In this way, with the difference of test method and fatigue judgment index, the healing index also had various measurement variables. The beam specimens, splitting specimens and cylindrical compression specimens can be used to conduct the self-healing test, and the deformation recovery rate, tensile strength ratio, and complex modulus ratio were obtained to become the healing index [49].

Researchers usually used fatigue–healing–fatigue (FHF) testing methods to assess the self-healing ability of bituminous materials [40, 62, 75–77], and the variation of complex modulus of bitumen was taken as the self-healing index. Figure 8 showed a typical plot of modulus decrease before and after self-healing in DSR experiments. In this context, Bommavarm et al. [75] carried out the DSR time scanning test to study the fatigue healing, and the ratio of complex modulus before and after self-healing was taken as self-healing index, see Eq. (2):

$$SHI = \frac{G_b^* - G_a^*}{G_0^*} \tag{2}$$

where SHI was the Self-healing index, and the meanings of other symbols were shown in the Fig. 8.

Additionally, Shan [78] found an effective self-healing index by a systematic comparative study, see Eq. (3):

$$SHI = \frac{G_b^*}{G_0^*} \times \frac{N_b}{N_a} \tag{3}$$

where SHI was the delf-healing index, and the meanings of other symbols were shown in the Fig. 8.

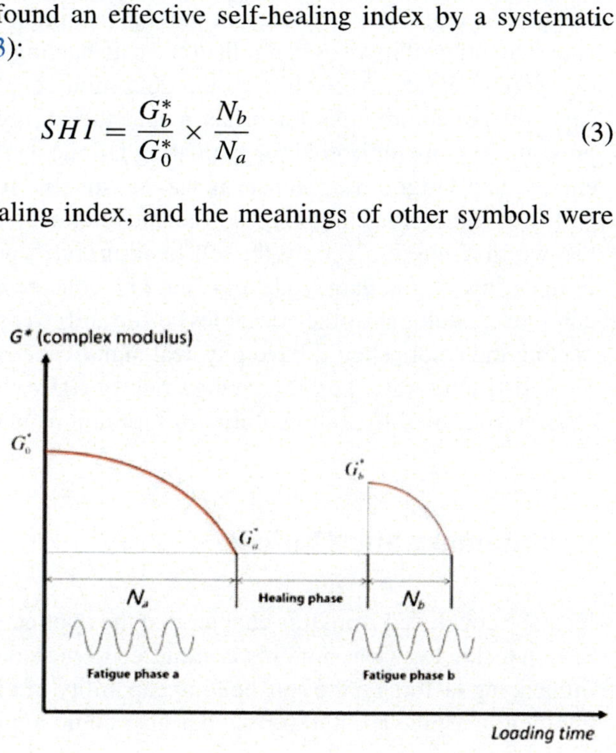

Fig. 8 Schematic diagram of complex modulus (G^*) decrease before (b) and after (a) self-healing

Moreover, Sun et al. [40] presented the damage ratio degree before and after self-healing as the self-healing index, see Eq. (4), and studied the self-healing potential of five kinds of bituminous mastics at 15 °C, 20 °C and 23 °C.

$$HI = \frac{G_b^* - G_a^*}{G_0^* - G_a^*} \quad (4)$$

Further, the self-healing activation energy of different materials was proposed as a reliable index to characterize the intrinsic self-healing ability of bituminous materials according to Arrhenius equation, see Eq. 5.

$$SHI(T, t) = SHI_0 + K_0 \exp\left(-\frac{E_h}{RT}\right) t^{0.25} \quad (5)$$

where SHI was the self-healing index, SHI_0 represented the instantaneous self-healing recovery, K_0 was the pre-exponential factor, R was the universal gas constant (8.314 J/mol/K), T was the Kelvin temperature (K). E_h was the self-healing activation energy, which represented the minimum energy required for the time-dependent strength gain.

Additionally, Shen et al. [78] used dissipated strain energy rate RDEC to evaluate the fatigue properties of bitumens. It was pointed out that the relationship curve between RDEC and loading times had a stable plateau interval. It was found that the plateau value PV had a logarithmic linear relationship with the self-healing time, and the slope of PV could measure the self-healing level of bitumen. In short, all the above-mentioned indexes can reflect the self-healing ability of bituminous materials to a certain extent. According to the definition of fatigue healing, the self-healing capability of bituminous materials meant the ability to resist fatigue damage. At present, the main indicators for evaluating fatigue were complex modulus, dissipated energy, and fatigue life. Hence, it was reasonable to characterize the self-healing ability of bituminous materials by the above indices, but the rationality of the value was worth discussing. Firstly, the self-healing ability was related to the damage degree of materials, so the damage level should be considered, not only the change rate of complex modulus and other variables before and after self-healing. Secondly, the self-healing indicators should have physical significance and differentiation. Moreover, the self-healing rate, i.e., the change rate of SHI with time, could also be used to characterize the self-healing ability of bituminous materials.

5 Summary and Outlook

It can be concluded from this chapter that the application of the multiscale characterization technology was not only committed to provide the diverse research results of influencing factors on the self-healing capability of bituminous materials (including bitumen, mastics, and mixtures), but also could supplement in-depth insights into comprehending the self-healing mechanism from various scales. For instance, the bitumen-filler system, bituminous mixtures, and bitumen-polymer system had the totally different self-healing properties with base bitumen. The multiscale analysis to the complex physical and chemical interactions between the base bitumen and other inclusions (such as fillers, aggregates, polymers, and so on) could help to reveal the self-healing mechanisms.

From the microscale point of view, the SEM and AFM techniques were usually applied to observe the micro-scale self-healing behavior in bitumen. However, the light bitumen components may volatilize or splash during electron beam scanning, which would seriously affect the observation results of bitumen microstructures. By contrast, AFM could be used to study the damage healing characteristics at the microstructure level, meanwhile it could probe the micro-scale mechanical properties of bitumen phases from nanometre to micrometer length scale. Moreover, some indirect observations by using microscopic characterization methods, such as FTIR, GPC, NMR, and TLC, also had significant roles in the self-healing research of bituminous materials. In this context, it can be concluded that the inherent self-healing capability of bitumen was closely related to its chemical components and molecular structures. At the meso-scale level, the self-healing process (i.e., crack closure) of bituminous materials under different conditions can be directly detected through the FM or X-ray CT techniques at a spatial and temporal scale. The adequate analysis to the FM or CT observation results was necessary to deeply excavate the meso-scale characteristics of self-healing behavior of bituminous materials.

The macro-scale self-healing measurements were basically mechanical related experiment methods, including fatigue-based self-healing tests, fracture-based self-healing tests, and field-based self-healing tests. Firstly, the fatigue-based self-healing tests were conducted with intermittent and interrupted load types or the combination of both. These tests revealed the self-healing ability of fatigue damage inside bituminous materials and could accurately simulate the self-healing process occurring in the bituminous pavements under the influences of repeated loads and environmental factors. Secondly, the fracture-based self-healing tests were performed to assess the self-healing responses by introducing the rest time and temperature in fractured bituminous specimens. These tests proved that the visible cracks could be gradually closed, and the mechanical properties can be recovered because of the inter-molecular wetting and diffusion between the crack faces. And thirdly, two kinds of field-based self-healing tests, FWD and SW test methods were introduced.

Finally, the concept of self-healing index (SHI) or self-healing rate were defined as the indicators to quantify the self-healing ability of bituminous materials. In this context, the SHI has different measurement variables depending on the test method and fatigue judgment index. Researchers usually use FHF testing approaches to assess the self-healing ability of bituminous materials. At present, the main indicators for evaluating fatigue were complex modulus, dissipated energy, and fatigue life. Therefore, it was reasonable to characterize the self-healing ability of bituminous materials by the SHI or self-healing rate. Recently, the linear amplitude scanning (LAS) test was developed to assess the fatigue property of bitumen. This test consumed far less time compared with FHF test. Hence, the corresponding self-healing test, i.e.,

LASH, was designed to assess the self-healing capability of bitumens [44]. However, the SHI based on the LASH required more studies to accurately compare the self-healing ability of various bituminous binders. All in all, it should be further studied for more precisely predicting the self-healing capability of bituminous materials by developing new evaluation methods and theories, so as to provide a solid theoretical and technical support for the design of long-life pavement structure and materials.

Acknowledgements The work described in this chapter was supported by National Natural Science Foundation of China (Nos. 52108390, 52178434, 51878500), State Key Laboratory of High Performance Civil Engineering Materials (No. 2021CEM010), Scientific Research Project of Beijing Educational Committee, China Postdoctoral Science Foundation (No. 2021M690271), Beijing Postdoctoral Research Foundation and Basic Research Project of Beijing University of Technology.

References

1. Yang W (1995) Macro-micro fracture mechanics. National Defense Industry Press, China
2. Qiu J (2012) Self healing of asphalt mixtures: towards a better understanding of the mechanism. Ph.D. Thesis, TU Delft University of Technology
3. Wool RP, O'Connor KM (1981) A theory crack healing in polymers. J App Phys 52:5953–5963
4. De Gennes PG (1971) Reptation of a polymer chain in the presence of fixed obstacles. J Chem Phys 55(2):572–579
5. Gaskin J (2013) On bitumen microstructure and the effects of crack healing. Ph.D. Thesis, University of Nottingham
6. Shen S, Sutharsan T (2011) Quantification of cohesive healing of asphalt binder and its impact factors based on dissipated energy analysis. Road Mater Pav Des 12(3):525–546
7. Roque R, Simms R, Chen Y, Koh C, Lopp G (2012) Development of a test method that will allow evaluation and quantification of the effects of healing on asphalt mixture. Florida Department of Transportation
8. Poulikakos LD, Partl MN (2012) A multi-scale fundamental investigation of moisture induced deterioration of porous asphalt concrete. Constr Build Mater 36:1025–1035
9. Lu X, Langton M, Olofsson P, Redelius P (2005) Wax morphology in bitumen. J Mater Sci 40:1893–1900
10. Shen S, Lu X, Liu L, Zhang C (2016) Investigation of the influence of crack width on healing properties of asphalt binders at multi-scale levels. Constr Build Mater 126:197–205
11. Peter E, Paul W (2010) Atomic force microscopy. Oxford University Press
12. Binnig G, Gerber C, Stoll E, Albrecht TR, Quate CF (1987) Atomic resolution with atomic force microscope. Europhy Let 3:1281
13. Blanchard CR (1996) Atomic force microscopy. Chem Edu 1:1–8
14. Raghavan D, Gu X, Nguyen T, VanLandingham M, Karim A (2000) Mapping polymer heterogeneity using atomic force microscopy phase imaging and nanoscale indentation. Macromolecules 33:2573–2583
15. Loeber L, Muller G, Morel J, Sutton O (1998) Bitumen in colloid science: a chemical, structural and rheological approach. Fuel 77:1443–1450
16. Claudy P, Letoffe JM, Rondelez F, Germanaud L, King G, Planche JP (1992) A new interpretation of time-dependent physical hardening in asphalt based on DSC and optical thermoanalysis. ACS symposium on chemistry and characterization of asphalts. Washington, DC
17. Nahar SN (2016) Phase-separation characteristics of bitumen and their relation to damage-healing. Ph.D. Thesis, TU Delft University of Technology

18. Kim YR, Little DN, Benson FC (1990) Chemical and mechanical evaluation on healing mechanism of asphalt concrete (with discussion). J Assoc Asph Pav Tech 59
19. Williams D, Little DN, Lytton RL, Kim YR, Kim Y (2001) Microdamage healing in asphalt and asphalt concrete, volume II: laboratory and field testing to assess and evaluate microdamage and microdamage healing. No. FHWA-RD-98-142
20. Santagata E, Baglieri O, Dalmazzo D, Tsantilis L (2009) Rheological and chemical investigation on the damage and healing properties of bituminous binders. Asph Pav Tech Proc 28:567
21. Sun D, Yu F, Li L, Lin T, Zhu XY (2017) Effect of chemical composition and structure of asphalt binders on self-healing. Constr Build Mater 133:495–501
22. Sun D, Sun G, Zhu X, Pang Q, Yu F, Lin T (2017) Identification of wetting and molecular diffusion stages during self-healing process of asphalt binder via fluorescence microscope. Constr Build Mater 132:230–239
23. Song I, Little DN, Masad EA, Lytton RL (2005) Comprehensive evaluation of damage in asphalt mastics using X-ray CT, continuum mechanics, and micromechanics (with discussion). J Assoc Asph Pav Tech 74
24. Tashman L, Masad E, Little D, Lytton RL (2004) Damage evolution in triaxial compression tests of HMA at high temperatures (with discussion). J Assoc Asph Pav Tech 73
25. Onifade I (2017) Development of energy-based damage and plasticity models for asphalt concrete mixtures. Ph.D. Thesis, KTH Royal Institute of Technology
26. Menozzi A, Garcia A, Partl MN, Tebaldi G, Schuetz P (2015) Induction healing of fatigue damage in asphalt test samples. Constr Build Mater 74:162–168
27. Garcia Á (2012) Self-healing of open cracks in asphalt mastic. Fuel 93:264–272
28. Norambuena-Contreras J, Garcia A (2016) Self-healing of asphalt mixture by microwave and induction heating. Mater Design 106:404–414
29. Norambuena-Contreras J, Serpell R, Valdés G, Gonzalez A, Schlangen E (2016) Effect of fibres addition on the physical and mechanical properties of asphalt mixtures with crack-healing purposes by microwave radiation. Constr Build Mater 127:369–382
30. Norambuena-Contreras J, Gonzalez-Torre I (2017) Influence of the microwave heating time on the self-healing properties of asphalt mixtures. App Sci 7(1076)
31. Gonzalez A, Norambuena-Contreras J, Storey L, Schlangen E (2018) Effect of RAP and fibers addition on asphalt mixtures with self-healing properties gained by microwave radiation heating. Contr Build Mater 159:164–174
32. Gonzalez A, Norambuena-Contreras J, Storey L, Schlangen E (2018) Self-healing properties of recycled asphalt mixtures containing metal waste: an approach through microwave radiation heating. J Environ Manage 214:242–251
33. Norambuena-Contreras J, Gonzalez A, Concha JL, Gonzalez-Torre I, Schlangen E (2018) Effect of metallic waste addition on the electrical, thermophysical and microwave crack-healing properties of asphalt mixtures. Constr Build Mater 187:1039–1050
34. Norambuena-Contreras J, Yalcin E, Garcia A, Al-Mansoori T, Yilmaz M, Hudson-Griffiths R (2018) Effect of mixing and ageing on the mechanical and self-healing properties of asphalt mixtures containing polymeric capsules. Constr Build Mater 175:254–266
35. Norambuena-Contreras J, Liu Q, Zhang L, Wu S, Yalcin E, Garcia A (2019) Influence of encapsulated sunflower oil on the mechanical and self-healing properties of dense-graded asphalt mixtures. Mater Struct 52:78
36. Norambuena-Contreras J, Yalcin E, Hudson-Griffiths R, Garcia A (2019) Mechanical and self-healing properties of stone mastic asphalt containing encapsulated rejuvenators. J Mater Civil Eng 31(5):04019052
37. Lu X, Soenen H, Redelius P (2003) Fatigue and healing characteristics of bitumens studied using dynamic shear rheometer. Proceedings 6th RILEM Symposium PTEBM vol 3. pp 408–415
38. Shen S, Chiu HM, Huang H (2010) Characterization of fatigue and healing in asphalt binders. J Mater Civ Eng 22:846–852

39. Van den Bergh W, Molenaar AAA, Van de Ven MFC, De Jonghe TC (2009) The influence of aged binder on the healing factor of asphalt mixtures. AESATEMA 2009 international conference on advances and trends in engineering materials and their applications, montreal, Canada
40. Sun D, Lin T, Zhu X, Cao L (2015) Calculation and evaluation of activation energy as a self-healing indication of bitumen mastic. Constr Build Mater 95:431–436
41. Castro M, Sánchez JA (2006) Fatigue and healing of asphalt mixtures: discriminate analysis of fatigue curves. J Trans Eng 132:168–174
42. Carpenter S, Shen S (2006) Dissipated energy approach to study hot-mix asphalt healing in fatigue. Trans Res Rec 1970:178–185
43. Jiang J, Ni F, Wu F, Sadek H, Lv Q (2019) Evaluation of the healing potential of asphalt mixtures based on a modified semi-circular bending test. Constr Build Mater 196:284–294
44. Xie W, Castorena C, Wang C, Kim YR (2017) A framework to characterize the healing potential of asphalt binder using the linear amplitude sweep test. Constr Build Mater 154:771–779
45. Li L, Gao Y, Zhang Y (2020) Crack length based healing characterisation of bitumen at different levels of cracking damage. J Clean Prod 258:120709
46. Bahia HU, Zhai H, Bonnetti K, Kose S (1999) Non-linear viscoelastic and fatigue properties of asphalt binders. J Assoc Asph Pav Tech 68:1–34
47. Smith BJ, Hesp SAM (2000) Crack pinning in asphalt mastic and concrete: effect of rest periods and polymer modifiers on the fatigue life. In: Proceedings of the papers submitted for review at 2nd Eurasphalt and Eurabitumen Congress, Barcelona Spain
48. Li R, Karki P, Hao P (2020) Fatigue and self-healing characterization of asphalt composites containing rock asphalts. Constr Build Mater 230:116835
49. Kim B, Roque R (2006) Evaluation of healing property of asphalt mixtures. Trans Res Rec 1970:84–91
50. Baaj H, Mikhailenko P, Almutairi H, Di Benedetto H (2018) Recovery of asphalt mixture stiffness during fatigue loading rest periods. Constr Build Mater 158:591–600
51. Abo-Qudais S, Suleiman A (2005) Monitoring fatigue famage and crack healing by ultrasound wave velocity. Nondestr Tes Eva 20:125–145
52. Little DN, Lytton RL, Chairl B, Williams D, Texas A (1999) An analysis of the mechanism of microdamage healing based on the application of micromechanics first principles of fracture and healing. J Assoc Asph Pav Tech 68
53. Lee HJ, Daniel JS, Kim YR (2000) Laboratory performance evaluation of modified asphalt mixtures for Inchon airport pavements. Inter J Pav Eng 1:151–169
54. Sun G, Sun D, Guarin A, Ma J, Chen F, Ghafooriroozbahany E (2019) Low temperature self-healing character of asphalt mixtures under different fatigue damage degrees. Constr Build Mater 223:870–882
55. Zhu X, Fan Y, Yu Y, Gilabert FA (2020) Crack propagation and microwave healing of Ferrite-filled asphalt mixes based on image correlation observations. Constr Buil Mater 262:119978
56. Ayar P, Moreno-Navarro F, Sol-Sánchez M, Rubio-Gámez MC (2018) Exploring the recovery of fatigue damage in bituminous mixtures: the role of rest periods. Mater Struc 51(1):1–10
57. Kim YR, Little DN, Lytton RL (2001) Use of dynamic mechanical analysis (DMA) to evaluate the fatigue and healing potential of asphalt binders in sand asphalt mixtures. Asphalt Paving Technol: Assoc Asphalt Paving Technol-Proc Tech Sessions 71:176–206
58. Kim YR, Little DN, Lytton RL (2003) Fatigue and healing characterization of asphalt mixtures. J Mater Civ Eng 15:75–83
59. Lytton RL, Uzan J, Fernando EG, Roque R, Hiltunen D, Stoffels SM (1993) Development and validation of performance prediction models and specifications for asphalt binders and paving mixes. Strategic Highway Research Program, Washington, DC
60. Si Z, Little DN, Lytton RL (2002) Characterization of microdamage and healing of asphalt concrete mixtures. J Mater Civ Eng 14(6):461–470
61. Mehrara A, Khodaii A (2016) Quantification of damage recovery of asphalt concrete as a consequence of rest time application using dissipated energy. Mater Stru 49:2947–2960

62. Sun G, Hu M, Sun D, Deng Y, Ma J, Lu T (2020) Temperature induced self-healing capability transition phenomenon of bitumens. Fuel 263:116698
63. Pronk AC, Cocurullo A (2009) Investigation of the PH model as a prediction tool in fatigue bending tests with rest periods. Advanced testing and characterisation of bituminous materials, Loizos, Partl, Taylor & Francis Group, London
64. Mamlouk M, Souliman M, Zeiada W (2012) Optimum testing conditions to measure HMA fatigue and healing using flexural bending test. TRB annual meeting No. 12-1266
65. Bazin P, Saunier JB (1967) Deformability, fatigue and healing properties of asphalt mixes. Intl Conf Struct Design Asphalt Pvmts Preprint 438–451
66. Jahanbakhsh H, Karimi MM, Nejad FM (2020) Correlation between asphalt concrete induced healing and rheological properties of asphalt binder. Constr Buil Mater 265:120577
67. Al-Mansoori T, Norambuena-Contreras J, Garcia A (2018) Effect of capsule addition and healing temperature on the self-healing potential of asphalt mixtures. Mater Stru 51(2):1–12
68. Xu S, Liu X, Tabaković A, Schlangen E (2020) A novel self-healing system: towards a sustainable porous asphalt. J Clean Prod 259:120815
69. Bhasin A, Little DN, Bommavaram R, Vasconcelos K (2008) A framework to quantify the effect of healing in bituminous materials using material properties. Road Mater Pav Des 9:219–242
70. Qiu J, Van de Ven M, Wu S, Yu J, Molenaar A (2012) Evaluating self healing capability of bituminous mastics. Exper mech 52:1163–1171
71. Lv Q, Huang W, Zhu X, Xiao F (2017) On the investigation of self-healing behavior of bitumen and its influencing factors. Mater Des 117:7–17
72. Zhou L, Huang W, Zhang Y, Lv Q, Yan C, Jiao Y (2020) Evaluation of the adhesion and healing properties of modified asphalt binders. Constr Buil Mater 251:119026
73. Kim Y, Kim YR (1997) In situ evaluation of fatigue damage growth and healing of asphalt concrete pavements using stress wave method. Trans Res Rec 1568:106–113
74. Groenendijk J (1998) Accelerated testing and surface cracking of asphalt concrete pavements. Ph.D. Thesis, Delft University of Technology
75. Bommavaram R, Bhasin A, Little DN (2009) Determining intrinsic healing properties of asphalt binders: role of dynamic shear rheometer. Trans Res Rec 2126:47–54
76. Sun G, Ma J, Sun D, Yu F (2021) Influence of weather accelerated ageing on healing temperature sensitivity of asphalts. J Clean Prod 281:124929
77. Sun G, Ma J, Sun D, Li B, Ling S, Lu T (2020) Influence of thermal oxidative aging on temperature induced self-healing transition of polymer modified bitumens. Mater Des 192:108717
78. Shan LY (2010) Fatigue and rheology mechanism of asphalt binder based on viscoelastic characteristic. Ph.D. Thesis, Harbin Institute of Technology

Numerical Simulation of Self-Healing Cementitious Materials

B. L. Freeman and A. D. Jefferson

1 Introduction

Significant progress has been made on the development of numerical models for SHCMs over the past two decades, and now a wide range of numerical approaches are available for simulating these materials. These models can play an important role in the development of SHCMs since, once calibrated, they can be used to investigate a wide range of parameters that would otherwise prove extremely time-consuming or impracticable using physical tests. In addition, models will also be necessary for the design and analysis of structures formed from SHCMs. The challenge in simulating SHCMs lies in the many interacting physical processes that govern their behaviour, all of which need to be captured adequately. These include mechanical processes, such as cracking and healing; transport processes, such as the flow of healing agents; chemical processes, including the curing of healing agents; and biological processes, such as bio-mineralisation. The specific physical mechanisms that need to be simulated depend upon the SHCM being considered. For instance, if the healing agent is a mineral solution, healing is achieved through the precipitation of minerals within the crack; whereas, if the agent is an adhesive, healing results from the chemical curing of the adhesive within the damaged zone of material.

In addition to reviewing models that were developed specifically for SHCMs, some related self-healing models for other materials are considered that simulate processes relevant to SHCMs.

The structure of the remainder of this chapter is as follows:

i. Sect. 2 details the simulation of mechanical processes;

B. L. Freeman (✉) · A. D. Jefferson
Cardiff University School of Engineering, The Parade, Cardiff C24 3AA, UK
e-mail: freemanbl@cardiff.ac.uk

A. D. Jefferson
e-mail: jeffersonad@cardiff.ac.uk

© Springer Nature Switzerland AG 2022

A. Kanellopoulos and J. Norambuena-Contreras (eds.), *Self-Healing Construction Materials*, Engineering Materials and Processes,
https://doi.org/10.1007/978-3-030-86880-2_6

ii. Sect. 3 details the simulation of transport processes;
iii. Sect. 4 presents a review of coupled models for the simulation of SHCMs;
iv. Sect. 5 presents a discussion of the challenges in modelling SHCMs;
v. Sect. 6 presents some concluding remarks.

2 Simulation of Mechanical Processes

2.1 Continuum Damage-Healing Mechanics

Continuum damage mechanics (CDM) provides a convenient framework from which to develop damage-healing models at the macroscale. It is necessary to introduce a few concepts from CDM theory before proceeding to explain how CDM models have been extended to simulate healing.

In the CDM framework, material degradation is simulated as a reduction in stiffness of the material, through the damage variable (ω) that represents distributed defects such as voids or micro-cracks. For an idealised isotropic material, the scalar damage variable is related to the area of defects in a representative plane of area, A (see Fig. 1):

$$\omega = \frac{A_\omega}{A} \tag{1}$$

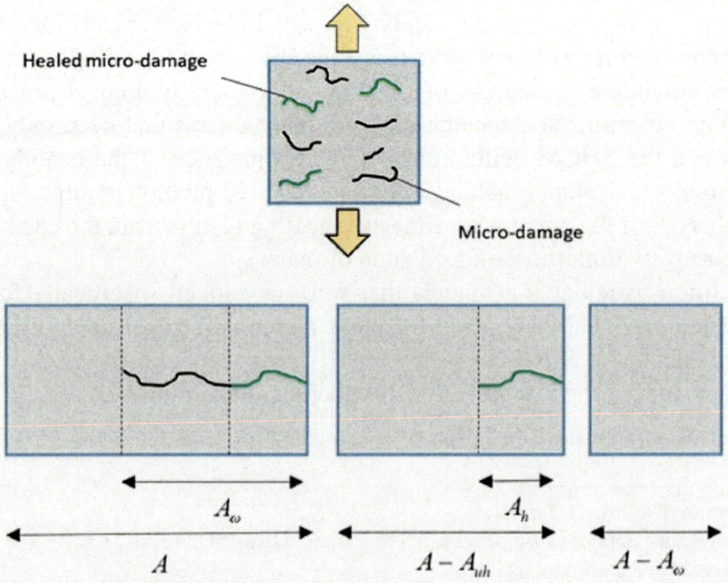

Fig. 1 Illustration of CDHM showing specimen in tension (top) and an equivalent material representation of the nominal state (left), the healing state (middle) and the effective state (right)

where A_ω is the area of defects intersecting the plane and by definition $\omega \in [0, 1]$.

For materials that may be assumed to damage in an isotropic manner, the constitutive relationship between the stress ($\boldsymbol{\sigma}$) and strain ($\boldsymbol{\varepsilon}$) tensors is given by:

$$\boldsymbol{\sigma} = (1 - \omega) \cdot \boldsymbol{D} : \boldsymbol{\varepsilon} \tag{2}$$

where $\boldsymbol{D}$ is the elasticity tensor.

A few examples of some widely referenced CDM models include those of Lemaitre [1], Mazars [2] and Simo and Ju [3].

A more general form of damage formulation employs a damage tensor that allows for different levels of damage in different directions [4]. Furthermore, multiple damage variables may be utilised to represent different degradation mechanisms [5]. For example, Comi and Perego [6] employed two scalar variables, one for compressive damage and the other for tensile damage, in their triaxial model for concrete.

CDM formulations require a damage evolution law. The form of this law varies greatly between models, with many employing a thermodynamic driving force, or energy release rate, to drive the damage, as follows:

$$Y = -\frac{\partial \Psi}{\partial \omega} = \frac{1}{2}\boldsymbol{\varepsilon} : \boldsymbol{D} : \boldsymbol{\varepsilon} \tag{3}$$

where Ψ is the Helmholtz type energy potential given as:

$$\Psi = \frac{1}{2}(1 - \omega)\boldsymbol{\varepsilon} : \boldsymbol{D} : \boldsymbol{\varepsilon} \tag{4}$$

and noting that $\partial \Psi / \partial \boldsymbol{\varepsilon} = \boldsymbol{\sigma}$.

The thermodynamic driving force is then used in a damage evolution equation, which, for cementitious materials such as concrete, often takes the form of an exponential softening function.

Damage only increases when the state of strain, stress or strain energy—depending on the formulation—is on the damage surface. A typical form for a damage surface (or loading function) can be expressed as:

$$F(Y, \kappa) = f(Y) - \kappa = 0 \tag{5}$$

where κ is a damage history variable and $f(Y)$ is a function of the thermodynamic driving force.

Finally, to complete the damage framework, the loading/unloading, or Kuhn–Tucker, conditions are required that are given as:

$$\dot{F} \leq 0, \quad \dot{\kappa} \geq 0, \quad F\dot{\kappa} = 0 \tag{6}$$

Continuum damage-healing mechanics (CDHM) is the natural extension of CDM to include healing effects. In CDHM, the constitutive relationship is extended to allow for the healing of the damaged component of material:

$$\boldsymbol{\sigma} = (1 - \omega) \cdot \boldsymbol{D} : \boldsymbol{\varepsilon} + h \cdot \omega \cdot \boldsymbol{D} : (\boldsymbol{\varepsilon} - \boldsymbol{\varepsilon}_h) \tag{7}$$

where h represents the effective area of healed material that is related to the damaged area:

$$h = \frac{A_h}{A_\omega} = \frac{A_\omega - A_{uh}}{A_\omega} \tag{8}$$

where $A_h \leq A_\omega$ and A_{uh} are the areas of healed material and unhealed material intersecting the representative plane, respectively, and by definition $h \in [0, 1]$.

An important feature of Eq. (7) is the inclusion of the healing strain ($\boldsymbol{\varepsilon}_h$) that is required to ensure thermodynamic consistency when healing occurs in a nonzero strain state.

More general forms of CDHM model frameworks employ healing tensors rather than a scalar healing variable. Furthermore, multiple healing variables can be introduced into the constitutive model in order to represent healing of different types of damage, or different healing mechanisms.

A number of authors employ the concept of effective and healing states in their formulations, in which the stresses are defined as (see, e.g., [7]):

$$\widehat{\boldsymbol{\sigma}} = \tfrac{\sigma}{1-\omega}, \ \widehat{\widehat{\boldsymbol{\sigma}}} = \tfrac{\sigma}{1-\omega_{eff}} \tag{9}$$

where $\widehat{\boldsymbol{\sigma}}$ and $\widehat{\widehat{\boldsymbol{\sigma}}}$ are the stresses in the effective and healing states, respectively, and $\omega_{eff} = \omega(1 - h)$ is an effective damage parameter. An equivalent material representation of the nominal, effective and healing states can be seen in Fig. 1.

Barbero et al. [8] presented a CDHM model for self-healing composite materials. The model was developed in a consistent thermodynamics framework that employed internal variables to represent the damage, plasticity and healing. The equations describing the evolution of the internal variables were derived from dissipation potentials. A particular feature of the approach was the proposition of a healing potential that was of the same form of the damage potential. From this, an associated healing function was derived that was the analogue to the damage function. This approach to simulating the healing was later adopted by a number of researchers in their models. A process for calibrating the healing parameters was described, which considers a shear test to failure and that gives the product of damage and healing as the ratio between the virgin and ultimate shear moduli. Finally, the model was applied to an example problem concerning a composite material loaded in shear. Following the calibration of the model with the data corresponding to a non-self-healing material, a parametric study was undertaken to demonstrate the effect of healing on the material response.

Granger et al. [9] presented an experimental and numerical study on healing behaviour in ultra-high performance concrete (UHPC). The authors presented a damage-healing constitutive model, derived from a thermodynamic potential. The resulting model reads:

$$\sigma = (1 - \omega)E\varepsilon + (1 - \omega_h)E_h(\Gamma_h)\varepsilon \tag{10}$$

in which E is Young's modulus and E_h is an effective Young's modulus of the healed material, which is a function of a curing parameter Γ_h. It is noted that the possibility of re-damage is accounted for in Eq. (10) through the inclusion of the multiplier $(1 - \omega_h)$ in the second term on the right-hand side, where ω_h is the damage variable that describes the proportion of healed material that has been damaged.

Upon examination of Eq. (10), two things can be noted; the first is that no healing strain is included in the second right-hand side term. This term is necessary when healing takes place under nonzero strain conditions. In the example problem presented, however, the specimen was unloaded during curing. In this case, the healing strain can be assumed to be zero and can be neglected. The second is that there is no term to account for the proportion of healed material, or degree of healing h. In this approach, this is instead taken into account in the $E_h(\Gamma_h)$ term, which implicitly accounts for the fact that only the damaged proportion of material undergoes healing.

The model predictions were compared to results from the experimental investigation, in which specimens were cracked in bending, before being submerged in water for different time periods, to determine the effect of curing time on the post-healed mechanical response. The model was found to be capable of representing the main features of the healing behaviour observed in the experiments.

Mergheim and Steinmann [10] proposed a model for self-healing polymers that allowed for simultaneous damage-healing, healing under nonzero strain conditions and healing of material in an initially stress-free state (i.e. it does not contain any elastic energy). The latter property means that the healing has no effect on the mechanical response of the material, until the strain state is modified. The proposed constitutive relationship is as follows:

$$\boldsymbol{\sigma} = (1 - \omega)\mathbf{D}_e : \boldsymbol{\varepsilon} + (1 - \omega_h) \int_{s=0}^{s=t} \frac{\mathrm{d}h}{\mathrm{d}s} \cdot \mathbf{D}_e : (\boldsymbol{\varepsilon}(t) - \boldsymbol{\varepsilon}(s))\, \mathrm{d}s \tag{11}$$

in which h represents the relative proportion of healed material and s is a time variable.

The simulation of the healing itself adopted an exponential curing function, which gives the degree of healing as a function of time as:

$$h(t) = \int_{s=t_c}^{t} \omega(s)\eta_h \exp(-\eta_h[t-s])ds \tag{12}$$

in which η_h is a curing parameter and t_c is the time that healing began.

In order to investigate the performance of the constitutive model, a parametric study was undertaken. The study investigated the response of the model to a range of different strain paths and curing times between loading cycles. The results of the study demonstrated interesting interactions between damage, healing and subsequent re-damage.

Abu Al-Rub and co-workers [7, 11–13] presented a series of CDHM models developed for the simulation of asphalt materials. In [7], the most comprehensive model was presented. The model was developed in a thermodynamics framework, a key feature of which was the decomposition of the conjugate forces into their energetic and dissipative components. The authors employed the concept of effective states in their model that was related through a power equivalence hypothesis. The constitutive model accounted for the temperature-dependent viscoelastic–viscoplastic–viscodamage behaviour and micro-damage healing of the material. The micro-damage healing evolution model employed was that proposed in [11] that reads:

$$\frac{\partial h}{\partial t} = \Gamma_h(1-\omega)^{-k_2}(1-h)^{k_1} \exp\left[-\theta_4\left(1-\frac{T}{T_0}\right)\right] \tag{13}$$

where Γ_h is a rate parameter, k_1 and k_2 are material parameters, θ_4 is a temperature healing coupling parameter and T_0 is a reference temperature.

A key feature of Eq. (13) is the inclusion of the effect of temperature through the Arrhenius-type term.

In the second linked paper [12], the model parameters were calibrated using experimental data, before a series of validation examples were considered. It was found that the model predictions were generally in good agreement with the experimental data, though there were discrepancies when simulating creep recovery at high temperature. Esgandi and El-Zein [14] presented a CDHM model for unsaturated geomaterials. The model was developed in a damage-healed framework using an energy equivalence hypothesis. The model employed an elastoplastic–damage–healing constitutive model to describe the material behaviour, where the healing rate was assumed constant. The model performance was demonstrated through a series of example problems, which showed that the inclusion of healing led to a more accurate prediction of the experimentally observed behaviour.

A model that allows for the simultaneous evolution of damage and healing in SHCMs was presented in [15]. A time-dependant exponential healing function was employed that was a function of damage and a material parameter that can be tuned to reflect material characteristics such as strength of the healed material or the volume fraction of microcapsules. A series of one-dimensional example problems were considered that studied the healing efficiency under various loading conditions, including uncoupled damage-healing evolution with zero and nonzero constant strain during the healing period, and the coupled damage-healing evolution. In addition to investigating the effect of the loading conditions, the examples were also used to study the effect of the healing period time and the healing rate constant. Oucif et al. [16] explored the concept of super-healing in CDHM (see also [17]). Super-healing is defined as the increase in strength of the material, beyond that of the original material. This phenomenon has been observed in experimental studies [18], where it was attributed to the strength of the healed zone been higher than that of the original material. The authors investigated differences in the analytical relationships between the variables in the different effective states, including a super-healed state, under both the equivalent elastic strain and equivalent elastic energy hypotheses for classical, generalised nonlinear and quadratic super-healing theories.

The theory of CDHM has been studied extensively by Voyiadjis and co-workers (see, e.g., [15–17, 19–23]). In [20], a viscoplastic-damage-healing model was derived in a thermodynamic framework. The model utilised a von Mises yield surface for the plasticity, and damage and healing surfaces that accounted for the isotropic and kinematic hardening effects. The formulation allowed for anisotropic damage and healing, for which two new healing tensors were proposed to describe the recovery of mechanical properties, including the damage and the elastic stiffness of the material. To demonstrate the capability of the model to capture realistic behaviour, experimental tests on self-healing SMP and PET specimens were simulated. The results of the simulations showed that the model was able to reproduce the behaviour observed during the experimental tests. Further work by the authors has included the concept of damageability and integrity of materials, and generalised nonlinear and quadratic healing [22], super-healing [17] and the decomposition of the healing tensor, to represent healing of cracks and voids [23].

2.2 Discrete Crack Mechanics

The CDHM framework represents the formation and healing of micro-cracks distributed in the continuum and as such is particularly well-suited to the simulation of SHCMs with embedded microcapsules. However, this modelling framework is less appropriate for simulating the behaviour of SHCMs that use embedded vascular networks for healing macro-cracks. For the simulation of these SHCMs, an explicit representation of the macro-cracks, including their location and geometry, is important. This is needed to allow the proper simulation of healing agent flow within the discrete cracks, and for the subsequent curing and healing processes within these discontinuities.

Whilst the techniques for incorporating discrete cracks into numerical models differ from each other [24], many employ cohesive zone constitutive models to relate the tractions across the crack to the crack opening displacements [25]. In this context, a preliminary study on modelling self-healing was undertaken by Schimmel and

Remmers [26] (See also [27]) who developed a 1D model to simulate idealised self-healing behaviour. The work employed a time-dependent healing function ($h_\omega(t)$) that was inserted into a one-D model:

$$\tau = (1 - \omega)K \mathrm{u} + (1 - \omega_h)\, \mathrm{r}\, h_\omega(t)\, \omega K\, (\mathrm{u} - u_h) \tag{14}$$

in which τ is the traction, K is the elastic stiffness, u is the crack opening, u_h is the crack opening at the time of healing and r is a model parameter that defines the limit of healing.

This preliminary work didn't include any validation against experimental data but did present a range of 1D example problems that explored the response of the material under a range of different loading conditions, in addition to a problem concerning a peeling test of a slender beam.

Maiti et al. [28] employed a cohesive zone model for the simulation of the mechanical behaviour of self-healing polymers. In their model, crack healing is simulated through the introduction of a wedge of healed material in the wake of the crack. When the thickness of this wedge is greater than or equal to the crack opening displacement, axial compressive forces, calculated using Lagrange multipliers, are transmitted across the crack, representing healing or crack retardation. A particular feature of the work is the use of a coarse-grained particle-based molecular dynamics simulation of the healing agent curing, that is subsequently related to the thickness of the artificial wedge. A series of fatigue tests are presented that show the effect of different parameters, including characteristic healing time and load amplitude, on the crack growth rate.

The thermodynamic framework presented in Abu Al-Rub and Darabi [7] and Darabi et al. [12, 13], was adapted by Alsheghri and Abu Al-Rub [29, 30] for a cohesive zone model. In their work, healing was assumed to take place at the crack-tip where the bond stress causes the crack faces to merge. The model separated damage and healing variables into components normal and tangential to the crack, but used the same evolution equations for both components. The degree of healing was calculated as the sum of an instantaneous contribution and a time-dependent term that was a function of the damage and healing history, crack closure and temperature. The model described in [30] was implemented in Abaqus and subsequently employed in a parametric study to determine the effect of a range of model parameters. Finally, a set of analyses were undertaken concerning crack-healing tests on PMMA specimens tested by Jud and Kausch [31]. The results showed that the model was able to reproduce the experimentally observed behaviour with good accuracy.

An interface model for simulating both autogenous and enhanced self-healing in cementitious materials was presented by Caggiano et al. [32]. Their model was based on a coupled damage-plasticity framework that was implemented using interface elements. Healing was taken into account through the modification of the parameters that defined the yielding/failure surface. To this end, a self-healing porosity factor, which evolves with the healing reaction, is employed to describe the restorative effects of healing. Following the calibration of model parameters, the model predictions were compared with the results of experimental tests undertaken by the same group. The tests considered cementitious specimens formed from both a standard concrete mix and a mix containing a crystalline admixture. The validation exercise showed that the model was capable of reproducing the experimental results with reasonable accuracy, including the enhanced healing of the specimens with the crystalline additive.

A model for self-healing quasi-brittle materials was presented by Zhang and Zhuang [33]. The model employed an element with an embedded strong discontinuity to represent the discrete crack, allowing for crack opening and crack sliding. Healing was introduced at the constitutive level in a mixed-mode traction separation law. The solidification of the healing agent was simulated using an Arrhenius-type law, whilst the relationship between the degree of healing and the regain in mechanical properties (namely strength and fracture energy) was assumed linear. The traction across a crack was calculated as the sum of the traction relating to the original and the healed material. The latter term was multiplied by a factor, α, introduced to take into account the level of penetration and contact between the healed and original materials. The effect of healing was demonstrated in example problems concerning a tension–shear test and a gravity dam. The results of the tension–shear test showed that healing can affect not only the load–displacement response, but also the crack pattern, as shown in Fig. 2.

A cohesive zone model that allows for multiple healing events is presented in Ponnusami et al. [34]. The model employs a bi-linear traction separation law, where the normal and shear tractions are given as weighted sum of the tractions due to the original and healed material, respectively. The weighting factors take into account an equivalent surface-based volume fraction of the original and healed material respectively. Upon healing, a shift is introduced into the traction separation law for the original material to account for the reduction in crack width as the crack is filled with healing agent. Healing of material in an initial stress-free state was accounted for by introducing a similar shift for the healed material. Healing is not explicitly calculated, though the framework can readily be coupled to a healing model. The performance of the model is investigated through a range of example problems, including healing under different loading conditions, multiple healing events and application to a particle-based self-healing system.

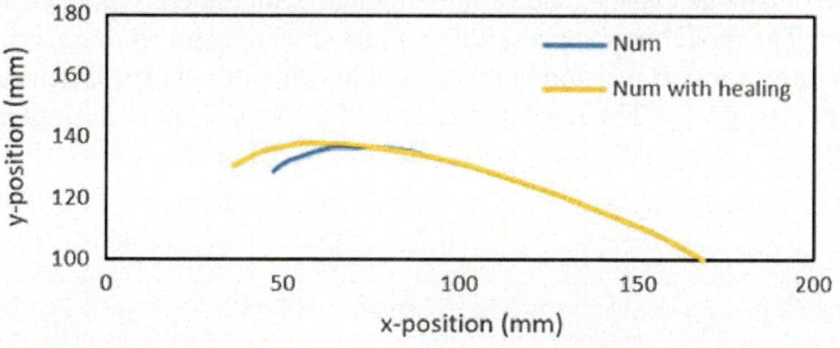

Fig. 2 Predicted final crack paths for the tension–shear test for healing and no healing case (after [33])

Xin et al. [35] presented a modelling framework for the simulation of the mechanics of a bio-mineralisation-based healing mechanism. The healed interface is simulated using a traction separation law, where the strength and stiffness of the cohesive zone are given as a function of the area fraction of calcium carbonate pillars that bridge the crack. The healing mechanism in this case was the bacteria-induced precipitation of calcium carbonate. The chemical reaction equation considered was given as:

$$\begin{aligned} &CO(NH_2)_2 + 2H_2O \xrightarrow{\text{Urease}} 2NH_4^+ + CO_3^{2-} \\ &Ca^{2+} + CO_3^{2-} \rightarrow CaCO_3 \end{aligned} \tag{15}$$

The model considered the nucleation of crystals that precipitate to form pillar-like structures, which grow in both length and radius. The growth model for an individual pillar was then used to derive a formula for cross-sectional area of bridged pillars, per unit interfacial area, that was subsequently employed in the mechanical model. The validation of the model was then carried out using their own experimental data on bio-mineralisation in ceramic and cement paste specimens. The model predictions compared well with the experimental data in terms of both the area fraction of calcium carbonate crystals and the healed strength ratio.

A cohesive zone model, applied to a finite element with embedded discontinuity, was proposed by Freeman et al. [24]. The model was based on the following the constitutive relationship:

$$\boldsymbol{\tau} = (1 - \omega) \cdot \boldsymbol{K} : \boldsymbol{u} + h \cdot \boldsymbol{K} : (\boldsymbol{u} - \boldsymbol{u}_h) \tag{16}$$

in which $\boldsymbol{\tau}$ denotes the tractions, $\boldsymbol{u}$ the crack opening displacements (or crack separation), $\boldsymbol{u}_h$ is the healing crack opening displacement vector and $h \in [0, \omega]$ denotes the proportion of healed material at the current time.

This model simulates the increase in the healing parameter with a rate-dependent exponential curing function and also allows for re-damage of the healed material. Thus, h can both decrease and increase depending on the amount of damage, healing, re-damage and re-healing. This model places no restrictions on the relative timings of healing and damage events and can thus simulate simultaneous damage and healing processes. The model was successfully validated using data from experiments on a series of direct tension tests on notched cuboid cementitious specimens undertaken by the authors' group. The results from two of these validation examples are given in Fig. 3.

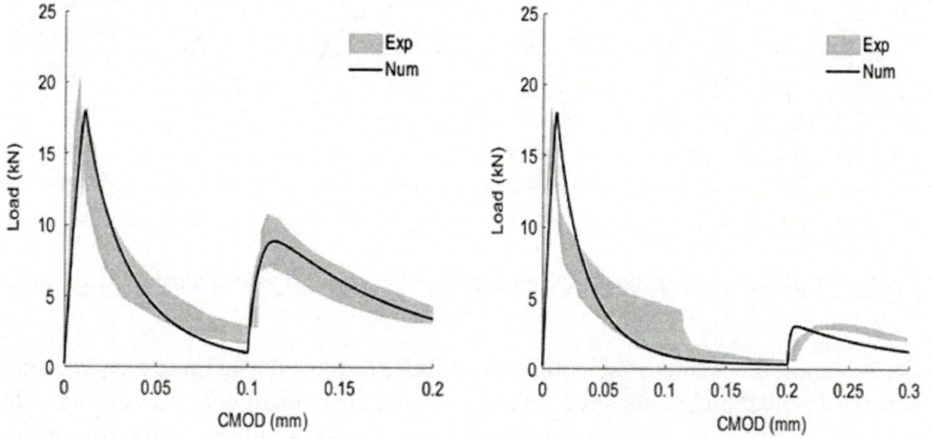

Fig. 3 Comparison of predicted mechanical response with experimental data (after [24])

2.3 *Micro/Mesoscale Mechanics*

Micromechanical and mesoscale models provide an attractive framework for simulating damage and healing at the microscale. Such models allow for a detailed description of the constitutive behaviour of SHCMs due to their intrinsic ability to represent anisotropic damage and healing. In addition to this, the effect of inclusions, such as microcapsules, on the behaviour of the composite material can be taken into account. Lower-scale models (i.e. scales below the macroscale) are characterised by their representation of the material as a multiphase composite. In the case of concrete, this composite usually consists of a mortar matrix that contains inclusions that represent the coarse aggregate particles. A key advantage of mesoscale models is their ability to directly simulate the physical processes occurring at the lower scales. An example of this is the rupture of microcapsules, which at the mesoscale can be simulated explicitly.

A micromechanical model was employed by Li et al. [36] to determine the effective properties of SHCMs containing microcapsules. The effects of the microcapsules and micro-cracks (both healed and unhealed) on the effective properties were accounted for using Eshelbian micromechanics [37]. An experimental programme that comprised mechanical tests, and sorption and vacuum saturation tests, supplemented the numerical investigation. The latter tests were used to determine the degree of healing. The comparison of the model predictions with the experimental data showed that the model could capture the effects of increasing the total volume of microcapsules in the cementitious matrix with good accuracy.

Quayum et al. [38] presented a preliminary study on the use of homogenisation techniques to determine the effective mechanical properties of self-healing cementitious composites. A range of homogenisation approaches was considered, including a Mori–Tanaka–Voigt scheme that considered three inclusions, namely the aggregate, capsule shell and the capsule core. In order to assess the accuracy of each of

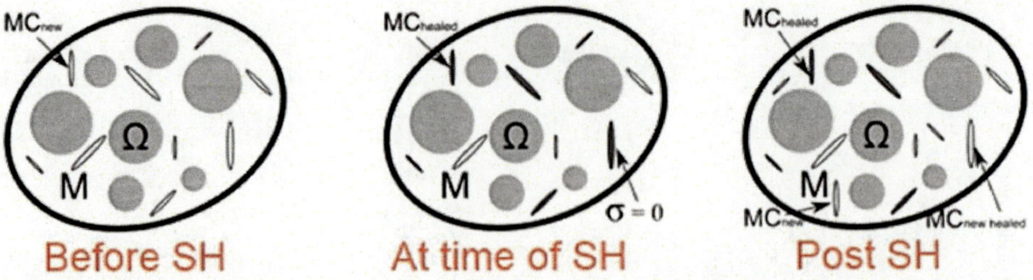

Fig. 4 Schematic representation of the micromechanical healing model (reproduced from [39])

the schemes, the results were compared to those produced by a finite element model that represented the inclusions explicitly. The authors concluded that care should be taken when applying linear homogenisation techniques to these materials and that further work should be undertaken to investigate the effects of the cargo properties on the response of the system.

Davies and Jefferson [39] developed a 3D micromechanical model for the simulation of SHCMs. Micro-cracking was accounted for through an added strain that is calculated as an integral of micro-crack strains around a hemisphere. This allows for an arbitrary degree of anisotropy. The healed micro-cracks were treated as inclusions that were assumed to heal in an initially stress-free state. This zero-stress change condition was ensured for nonzero strain states by the incorporation of a healing strain into the constitutive equation. In addition to this, the healed micro-cracks were allowed to re-damage (once). A schematic representation of the model can be seen in Fig. 4. The model was validated using data from experimental tests concerning autogenous healing of cementitious specimens.

An alternative approach to representing the meso-structure is to use a discrete element method (DEM). Herbst and Luding [40] employed a DEM to investigate self-healing particulate materials. In their model, contact laws between particles were history-dependent, and elasto-plasticity, viscosity, adhesion and friction were all considered. Healing was introduced as the repair of bonds that had been damaged. Healing was initiated immediately that damage was detected and was assumed instantaneous. The approach was implemented in a two-step algorithm that i) detected then healed damage and ii) ran the mechanical simulation for a given time, τ_d, before repeating. To demonstrate the effect of healing, a series of compression tests were considered. The results of the simulations showed that the healing adhesion parameter, and the rate of damage detection (defined as $1/\tau_d$) had a significant effect on the predicted mechanical response.

Zhou et al. [41] developed a new damage-healing model for SHCMs with embedded microcapsules. The model was implemented in the 3D DEM code PFC3D. In their model, the microcapsules are simulated implicitly, through the healing of the broken bonds. Healing is introduced in a similar manner to that of Herbst and Luding [40], whereby new bonds with properties pertaining to the healing agent are introduced to replace damaged bonds. In their work, only damaged bonds meeting a maximum opening criterion are healed, whilst the healing itself is assumed to take place whilst the specimen is unloaded. The effect of the microcapsules is taken into account in the expression for the maximum stresses that the healed bonds can withstand. The authors propose a healing index that is given as a function of the number of initial, healed and broken bonds (N_i, N_h and N_d, respectively), as follows:

$$h = \left[\frac{N_h - N_d}{N_i - N_d} \right] \tag{17}$$

The model is subsequently employed in a parametric study concerning the compressive loading of a concrete specimen. The study investigated the effect that the strength and stiffness of the healed material, level of initial damage and the effect of partial healing on the mechanical response. It was found that the mechanical properties of the healed material had a significant effect on the predicted response.

Gilabert et al. [42] developed a finite element model to simulate the mechanical behaviour of SHCMs with encapsulated healing agents. The model employed a combination of the extended finite element (XFEM) technique, to describe the fracture in the matrix and capsule itself, and the cohesive surface technique to describe the behaviour of the interface between them. The model was applied to two different self-healing applications. The first concerned the three-point bending of a concrete beam with embedded tubular capsules, whilst the second concerned a self-healing polymer containing a microcapsule. A comprehensive study is undertaken into the effect of a range of parameters, including the size of the capsules, the number of capsules and the strength of the interface between the capsules and the matrix in which they are embedded. A key finding that arose from the simulations was that a glass capsule with a radius of 1.5 mm and thickness 0.18 mm requires a minimum bond strength of 2 MPa in order to be fractured under mechanical loading. The authors state that the motivation behind the study was to show how such models could provide insights into the interaction between cracks and embedded capsules, which can be useful for experimentalists.

In their study, Šavija et al. [43] used a lattice model to investigate the breakage of polymeric tubular capsules embedded in concrete. In the lattice model, the concrete matrix and embedded capsules were discretised by a set of beam elements. Cracking was simulated by removing any elements that met a given fracture criterion. The elements representing the concrete matrix adopted a brittle fracture law, whilst those representing the embedded capsules adopted a multi-linear stress–strain relationship that was determined from experimental data. A perfect bond was assumed between the matrix and the capsules. Simulations were presented that concerned the uniaxial loading of a mortar block containing an embedded capsule, with capsule material and wall thickness being varied. A key finding of the work was that one type of PMMA capsules with wall thicknesses of 0.3 and 0.7 mm was suitable for application, but that PMMA capsules with greater wall thicknesses, and capsules made from different types of PMMA or polylactic acid were unsuitable due to the large crack widths observed at capsule rupture.

The mechanical properties of the interface between the healing agent and the cementitious matrix were investigated in a combined experimental and numerical

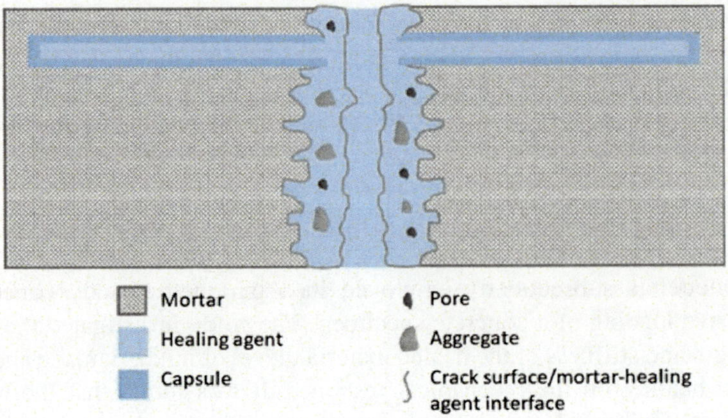

Fig. 5 Schematic of the interface between the healing agent and mortar, aggregates and pores (after [44])

study in Xue et al. [44]. The numerical model comprised four phases, representing the aggregate particles, mortar, healing agent and pores. The microstructure was generated using a Python program based on the random distribution of aggregate particles and pores within the matrix, before the healing agent was added. A cohesive surface model was used to simulate the mechanical behaviour that assumed the fracture takes place at the interfaces between the healing agent and the mortar, aggregates and pores, respectively (see Fig. 5).

The model was subsequently employed to investigate the interface stress distribution and load capacity, before the effects of changing the proportion and properties of the aggregate, pores and interface were explored. The results showed that high stress concentrations occur adjacent to aggregate particles and that fracture was initiated in these locations. The study also found that both aggregate particles and pores delay the propagation of interface cracks.

Mauludin et al. [45] (see also [46, 47]) investigated the crack behaviour of encapsulation-based self-healing concrete. Their model employed a cohesive zone approach to simulate the interaction between the matrix material and the embedded microcapsules. The model was implemented in Abaqus, and cohesive elements were inserted between all element edges to allow for the arbitrary propagation of the cracks. A series of numerical experiments were undertaken to determine the effect of varying fracture properties for the matrix, interfacial transition zone and microcapsules, and the core–shell thickness ratio. The results showed that the strength of the interfacial transition zone between the capsule and the matrix has a significant effect on the fracture propagation, and that a strength similar to that of the matrix leads to a greater chance of the capsule being ruptured. An example of this behaviour for different strengths of the interfacial transition zone can be seen in Fig. 6.

A similar investigation was undertaken in Krishnasamy et al. [48] (see also [49] for the mechanical behaviour of a self-healing composite), who considered the fracture behaviour of a self-healing thermal barrier coating. The parameters considered

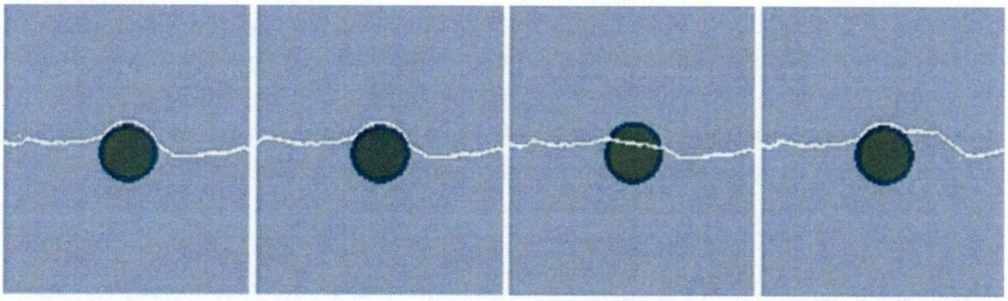

Fig. 6 Crack pattern of microcapsule with core–shell thickness ratio of 15:1 for interfacial zone strengths of (left) 25%, centre (left) 50%, centre (right) 75% and (right) 100% of the strength of the mortar matrix (reproduced from [47], with permission)

included a mismatch in the coefficient of thermal expansion and the relative strength and strength of the interface between the embedded healing particles and the matrix material. It was found that the mismatch in the coefficients of thermal expansion not only had a significant impact on the formation and growth of cracks, but also on the crack pattern, whilst varying the relative strength and strength of the interface between the particles and the matrix was found to have a significant effect on the failure of the composite material.

2.4 *Specific Aspects*

2.4.1 Probability of Crack Hitting an Embedded Microcapsule

The self-healing of SHCMs that utilise embedded microcapsules depends on the crack hitting, and subsequently fracturing, the embedded microcapsule. With this in mind, and in order to determine the dosage, size and shape of the microcapsules, a number of authors have investigated the probability of a crack hitting a microcapsule.

Zemskov et al. [50] presented analytical models for the probability of a crack hitting a capsule. The first model considered the random placement of capsules in a layered structure, whilst the second considered the entirely random placement of capsules. The models considered a range of parameters, including crack depth and capsule radius.

Huang and Ye [51] studied the effects of encapsulated water on the self-healing efficiency due to further hydration, using Hymostruc3D. As part of the study, a numerical investigation was undertaken to determine the probability of a crack hitting a capsule for a range of capsule sizes. To this end, a thousand numerical experiments, that used different random distributions of capsules, were carried out and the average number of capsules intersected by the crack was calculated, along with the amount of water released as a function of capsule dosage.

The optimal dosage of capsules was investigated using a probabilistic analysis in Wang et al. [52]. In the study, irregular concrete cracks were represented as hexagonal prisms, and both cylindrical and spherical capsules were considered. The intersection of cracks with the capsules was calculated using integral geometry theory, whilst the optimal dosage was calculated using the binomial distribution principle. The resulting theoretical models compared well to a computer simulation that considered the hexagonal crack pattern and randomly distributed capsules.

2.4.2 Numerical Simulation of Self-Healing System that Uses Shape Memory Polymer (SMP) Tendons

Hazelwood et al. [53] presented a model for the simulation of a self-healing system that uses SMP tendons. The model coupled a layered beam finite element model that accounted for hydration, creep and shrinkage in the cementitious matrix, to a phenomenological model for simulating the thermo-mechanical rheological behaviour of the SMP tendons [54, 55]. The model was validated using experimental data before being used to compare 10-year predictions of the crack widths of a reinforced concrete beam, with and without the SMP tendons. The simulations show that the crack widths could be reduced by 65% if the cracks are 50% healed and the tendons have a shrinkage stress potential of 100 MPa.

A beam-hinge model was employed by Balzano et al. [56] to simulate a self-healing system that uses hybrid SMP tendons. The model used a layered beam approach in the hinge zone and accounted for cracking of the cementitious matrix, as well as slippage of the tendons at the anchorage, through the introduction of a slip–displacement term. Application of the model to the simulation of the self-healing hybrid-SMP system showed that was able to reproduce the experimental behaviour with reasonable accuracy. In addition to this, the model proved to be a useful tool in the interpretation of the experimental results.

3 Simulation of Transport Processes

3.1 Matrix Transport

Transport processes play an important role in the self-healing mechanisms of SHCMs. For instance, the precipitation of calcite in a concrete crack depends upon the presence of carbon dioxide, calcium hydroxide and moisture, the concentrations of which directly affect the rate of the reaction. The simulation of their transport to (and from) a damage site is therefore of paramount importance for accurately predicting healing (or sealing). Another interesting example of the importance of transport processes can be found in the simulation of pH-sensitive (or chemically triggered) microcapsules [57]. In this case, the simulation of chemical transport is required not only for the calculation of the rates of the healing reaction, but also to predict the release of the healing agent in the first place.

The majority of porous media transport models are described by macroscopic balance equations that have been derived using hybrid mixture theory [58]. The porous medium is treated as a multiphase mixture, consisting of the solid skeleton and pores that contain the liquid and gaseous phases. In addition to this, the liquid phase may contain dissolved chemical species, whilst the gaseous phase is usually composed of a mixture of dry air and moisture vapour.

Assuming that the mechanical deformation does not significantly affect the transport behaviour, the macroscopic balance equations are given in the following, where n denotes the porosity, S_π and ρ_π represent the degree of pore saturation and density of the phase π, whilst J_π^a, J_π^d and $\dot{m}_\pi$ represent advective and diffusive fluxes and net production respectively.

(i) Mass balance of the liquid phase is given by:

$$\frac{\partial(nS_l\rho_l)}{\partial t} + \nabla \cdot J_l^a + \dot{m}_1 = 0 \tag{18}$$

(ii) Mass balance of the gas phase considers both dry air (Eq. 19) and moisture vapour (Eq. 20) and is given by:

$$\frac{\partial\left(nS_g\rho_a\right)}{\partial t} + \nabla \cdot J_a^a + \nabla \cdot J_a^d = 0 \tag{19}$$

$$\frac{\partial\left(nS_g\rho_v\right)}{\partial t} + \nabla \cdot J_v^a + \nabla \cdot J_v^d + \dot{m}_v = 0 \tag{20}$$

(iii) Mass balance of dissolved chemical species is given by:

$$\frac{\partial(c_i nS_l\rho_l)}{\partial t} + \nabla \cdot J_c^a + \nabla \cdot J_c^d + \dot{m}_c^i = 0 \tag{21}$$

where c_i denotes the concentration of species i.

It is noted that the mass balance of chemical species in the gas phase (for instance $CO_2(g)$ that may be considered in a carbonation model) is defined similarly where the liquid variables are replaced by their gas phase equivalents.

(iv) Mass balance of solid phases is given by:

$$\frac{\partial\left[c_s^i(1-n)\rho_s\right]}{\partial t} + \dot{m}_s^i = 0 \tag{22}$$

where c_s^i denotes the concentration of the solid species i, in the solid phase. It is noted that the form of Eq. (22) is suitable for considering the solid species individually. An alternative is to consider the mass balance of the solid phase as a whole, in which case Eq. (22) simplifies to: $\partial[(1-n)\rho_s]/\partial t + \dot{m}_s = 0$, where $\dot{m}_s$ denotes the net production of all solid species.

(v) Energy balance of the mixture is given by:

$$(\rho C_p)_{eff}\frac{\partial T}{\partial t} + (\rho_l C_p^l v_l + \rho_g C_p^g v_g)\nabla T + \nabla \cdot J_T^d + \dot{m}_T = 0 \tag{23}$$

where $(\rho C_p)_{eff}$ denotes the effective thermal capacity of the multiphase material, C_p^π is the isobaric specific heat of phase π and v_π is the velocity of the phase π.

Darcy's law is commonly employed to describe the advective fluxes, whilst Fick's law is used to describe the diffusive fluxes (with the exception of the energy balance equation where Fourier's law is employed). The exact form of the fluxes depends upon both the model and the application. For instance, the flux of the liquid moisture may include a term representing osmotic flow, which represents flow due to gradients in dissolved chemical species concentrations [59], in addition to the usual flow due to the pressure gradient. For certain applications, this term is necessary, whilst for others it may prove insignificant and can therefore be neglected. It is also noted that the equations above may be supplemented with the balance of linear momentum to describe the mechanical behaviour of the medium.

A hygro-thermo-chemical model for the simulation of the early age autogenous healing of cementitious materials were presented by Chitez and Jefferson [60]. The model was derived in the framework of hybrid mixture theory and considered the balance equations for moisture, unreacted cement and enthalpy. The model accounted for different forms of moisture within the CSH gel using the colloid model of Jennings [61]. The STOICH_HC2 model, proposed in Chitez and Jefferson [62], was used to trace the evolution of the pore structure. The STOICH_HC2 model considers the stoichiometry and hydration degree of the different unhydrated compounds, naturally allowing for different cement types and mixes and was derived to be consistent with the moisture mass balance equation. A schematic of the phases and material movements considered can be seen in Fig. 7.

The precipitation of hydrates was modelled as an equilibrium reaction using a Freundlich-type isotherm that relates the mass of precipitated hydrates to the concentration of the unreacted cement:

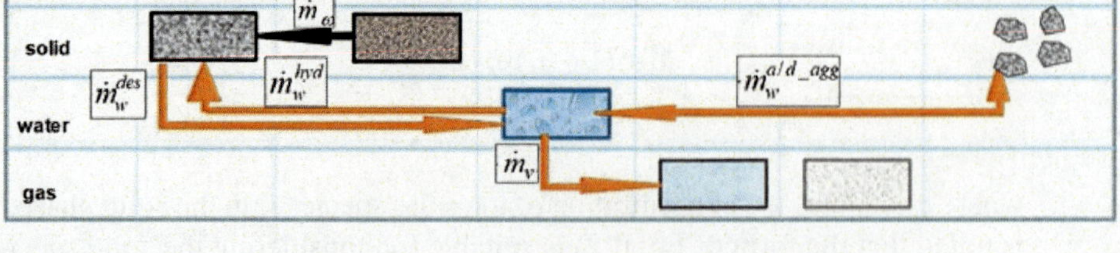

Fig. 7 Phases and material movements considered in the model (after [60])

$$m_h = \left(\alpha_p S_l^{cap} c_i^{\beta_p}\right) \rho_l n_{cap} \tag{24}$$

where m_h is the mass of precipitated healing materials, S_l^{cap} and n_{cap} are the degree of saturation of the capillaries and the capillary porosity, respectively, and α_p and β_p are reaction parameters.

The model predictions correlated well with experimental observations for different crack widths (from 10 μm up to 188 μm) and moisture conditions.

Carbonation reactions play a crucial role in the durability of concrete structures in terms of both the positive effects of autogenous self-healing but also the negative effects associated with reinforcement corrosion. Earlier work on the simulation of carbonation [63–66] focussed on the progress of the carbonation front in order to assess the risk of reinforcement corrosion. In contrast, more recent work [67–70] has instead focussed on the simulation of natural and enhanced carbonation for the self-healing of cracks.

Aliko-Benítez et al., [68] presented a chemical-diffusive model for the simulation of the self-healing of concrete through carbonation. The model considered the diffusive transport of CO_3^{2-} ions from the environment and into the specimen, whereupon they react with Ca^{2+} ions, assumed to be dispersed throughout the matrix, to form calcium carbonate. The reaction was calculated using a first-order kinetic equation given as:

$$\frac{\partial[\mathrm{CaCO_3}]}{\partial t} = k\left[\mathrm{CO_3^{2-}}\right] \cdot \mathrm{H}\left[\mathrm{Ca^{2+}}\right] \tag{25}$$

where $[i]$ represents the concentration of species i, k is a rate constant and H is the Heaviside function.

In order to simulate the healing of damage, the authors introduced damage and healing variables, in the spirit of CDHM (see Eqs. (1) and (8)). The healing was calculated as the ratio of the volume of precipitated material to that of the damage, whilst the damage parameter was employed as an internal variable that could be prescribed. The authors went on to present a study to determine the effects of a number of parameters, including the diffusion coefficient, reaction rate and level of damage on the healing of prismatic and T-beam specimens.

Javierre et al., [70] presented a reactive transport model for the simulation of carbonation reactions in self-healing concrete. The carbonation reaction considered the dissolution of gaseous carbon dioxide and solid calcium hydroxide (from the cement matrix), the carbonation of calcium hydroxide taking place in the moisture phase, and finally, upon solution saturation, the precipitation of solid calcium carbonate. The effect of the reaction on the pore structure was taken into account following the approach of Papadakis et al. [63]. The reactive chemical transport model was coupled to Richard's equation, which describes the movement of moisture within the cementitious matrix, in a staggered solution method. Some preliminary results are presented concerning the accelerated carbonation of a concrete specimen. The authors note that the comparison of the model predictions with experimental data is the subject of future research.

3.2 Discrete Crack Transport

For the simulation of transport in discrete cracks, the governing mass balance equations are of a similar form of those presented above. A key difference, for a number of models, lies in the calculation of the advective flux for the liquid and gas phases. A number of different approaches have been employed, many of which (including Darcy's law) can be derived from the Navier–Stokes momentum balance equation that reads:

$$\frac{D\mathbf{v}_\pi}{Dt} = -\frac{\nabla P_\pi}{\rho_\pi} + \mu_\pi \nabla^2 \mathbf{v}_\pi \tag{26}$$

where D/Dt denotes the material derivative and P_π and μ_π represent the pressure and dynamic viscosity of the phase π, respectively.

A model for the preferential moisture uptake through cracked building materials was presented by Carmeliet and co-workers [71–73]. The flow in the cracks was calculated using a quasi-static mass balance equation, coupled to Darcy's law that describes the movement of the moisture front. The flow in the cracks was coupled to the unsaturated matrix flow through a sink/source term, representing flow through the crack faces. In [72], the moisture transport model is coupled with a partition-of-unity crack model to investigate the effect of moisture on damage processes. The model predictions compared well with experimental measurements concerning the preferential moisture uptake in both naturally and artificially cracked bricks.

Hall et al. [74] presented a model for the simulation of healing agent flow in self-healing composite materials with embedded vascular networks. The model utilises a smooth particle hydrodynamics framework to solve the Navier–Stokes equations that describe the flow. Particular attention is paid to the treatment of the viscous force term due to its significance in the simulation of self-healing systems where the Reynolds number is typically small. To explore the capability of the model to predict crack filling, an example problem was presented that considered two delaminations, connected by a shear crack. The source of healing agent was a reservoir connected to the crack system and pressurised at 1.19×10^5 Pa. The simulations included an investigation of the effect of viscosity on the filling fraction. An illustration of the crack filling with a healing agent can be seen in Fig. 8. It was found that a higher viscosity leads to a smaller filling fraction for a given elapsed time.

The simulation of healing agent flow in discrete cracks was explored by Gardner et al. [75–77]. These models used a modified Lucas–Washburn equation that allowed for stick-slip of the meniscus, frictional dissipation, and wall slip. A range of fluids was considered, including plain water, ground-granulated blast-furnace slag (GGBS) and pulverised fly ash (PFA) solutions, and cyanoacrylate. The model predictions were shown to be able to accurately reproduce the observed capillary rise behaviour for a number of different crack configurations. Gardner et al. [76, 77] also investigated the effect of healing agent curing on the flow properties of the agents considered; for example, the viscosity of cyanoacrylate was measured for a range of curing

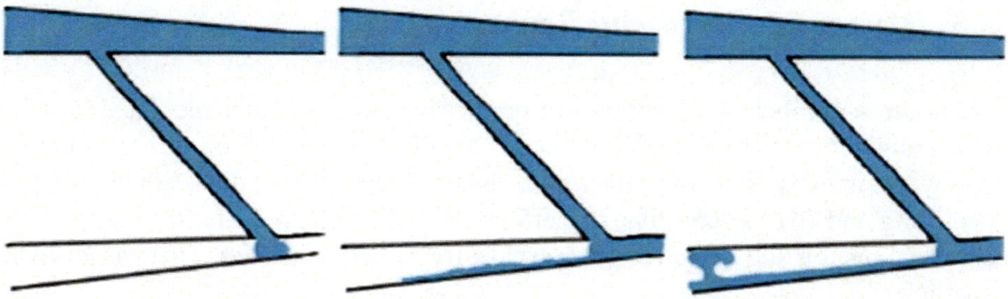

Fig. 8 Illustration of filling concept of delamination and a shear crack (after [74])

times (relative to the time the agent was first decanted and placed in the system) using a bespoke manometer. It was concluded that the viscosity of the cyanoacrylate remained unchanged over a time period of 15 min for the 4 mm diameter capillary channels used; however, the viscosity had increased appreciably two hours after the start of the experiments.

Gilabert et al. [78] simulated the flow of an autonomic healing agent in an encapsulation-based self-healing system, using the computational fluid dynamics code OpenFOAM [79]. The flow was described using the Navier–Stokes equations, whilst a piecewise linear dynamic contact angle relationship (given as a function of time) was employed that was calibrated using their own experimental data. The simulation considered the flow in both the crack plane and in the interior of the embedded capsules. The final leakage pattern predicted by the model was found to compare favourably with the experimental data obtained for a crack width of 300 μm.

An embedded discontinuity model for simulating the healing of cracks through calcium carbonate precipitation was presented in Ranaivomanana and Benkemoun [80]. The model considers the reactive transport of Ca^{2+} ions and CO_3^{2-} ions in the crack, coupled with reactive transport of Ca^{2+} ions in the matrix. The equations are coupled through a mass transfer term that represents the transport of ions through the crack face. The formation of calcium carbonate is calculated using a kinetic reaction model and its resultant effect on the crack width is accounted for through the consideration of the thickness of the layer of precipitated calcium carbonate on the crack face.

The sealing of cracks as a result of calcium carbonate precipitation was considered in detail by Iyer et al. [81], who developed a model for the simulation of transport through leakage pathways in wellbore cement. The model considered the two-phase flow of carbon dioxide and brine through a discrete crack. In addition to this, the reactive transport of calcium ions within the brine was also considered, whilst the dissolved carbon dioxide concentration was assumed constant at its solubility. The model employed a previously developed chemical reaction model [82–84], that considers dissolution of calcium hydroxide, along with the precipitation and dissolution of calcium carbonate. A series of 1D example problems were presented that demonstrate the effect of a range of parameters, including initial crack width, on the degree of crack-sealing.

3.3 Microscale/Mesoscale Transport

There are a number of advantages to employing microscale or mesoscale models, as highlighted in Sect. 2 of this chapter. Models at this scale are utilised in one of two ways, namely (i) to investigate the physical processes at the scale that they are occurring and (ii) to determine the effect of various phenomena, such as cracking and healing, on the effective macroscale transport properties, such as diffusion coefficients.

A number of authors have employed the Hymostruc model, originally developed by van Breugel [85], to study self-healing behaviour in cementitious materials (see, e.g., [86–89]). The Hymostruc model was developed to simulate cement hydration and the subsequent evolution of the microstructure. The first of these studies was undertaken by ter Heide [86], who employed Hymostruc, along with the bar and ribbon models of Koenders [90] and Lokhorst [91], to simulate autogenous healing. In their approach, sets of horizontal bars, which were linked by vertical elements, were used to represent clusters of hydration products. Ye and van Breugel [87] investigated the effectiveness of a 3D version of Hymostruc (Hymostruc3D, first developed by Koenders [90]), to simulate the healing of micro-cracks. The simulations showed the potential of many cementitious materials to heal micro-cracks in specimens, or structural elements that are up to a year old.

Koenders et al. [92] (see also [89]) used Hymostruc to investigate the healing potential of a SHCM employing dissoluble encapsulated particles (DEP). The healing mechanism was the hydration of the DEP particles, initiated by a crack but dependent on sufficient moisture being present at the location. A schematic that illustrates the healing mechanism is given in Fig. 9. The simulations reported by Koenders et al. [89, 92] employed a cubic domain of $100 \times 100 \times 100\,\mu m$, into which a $30\,\mu m$ crack was introduced. The study investigated the effect of changing a range of parameters on the properties of the cubic domain. The parameters considered included the percentage and grading fractions of cement replaced by DEP particles, the relative expansion of the DEP particles with respect to the cement particles, and (in [89]), the crack width. An example of the results produced by the model can be seen in Fig. 10. The results of the study led to some interesting conclusions:

(i) DEP systems can be modelled well with microstructural hydration models like Hymostruc.
(ii) Both the grading and expansion potential of DEP significantly affect the healing potential.
(iii) DEPs can act as a delayed source of hydration that can enable the self-healing of cracks.
(iv) The degree of hardening of a cementitious material system containing DEPs can reach the same levels as the original cementitious material.
(v) Whilst the introduction of DEPs leads to an initially higher porosity, the porosity of the system following the activation of DEPs is comparable to that of the original material.

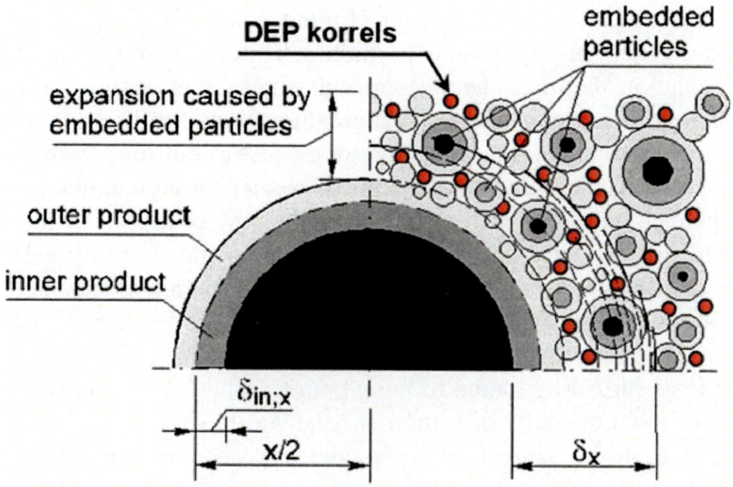

Fig. 9 Schematic representation of the microstructure of a hardening cement grain by means of the grain expansion mechanism as used in Hymostruc. The cement matrix is composed of hardening cement particles and unhydrated DEP grains that will be activated after cracking and moistening. (reproduced from [94], with permission)

In Koenders et al. [92] (see also [93]) preliminary work on the coupling Hymostruct3D to a mechanical model is presented. In this regard, the microstructure is analysed and converted to a voxel-based finite element mesh that can then be used in a mechanical model (e.g. that of Caggiano et al. [32]), to analyse the effect of healing on the mechanical properties.

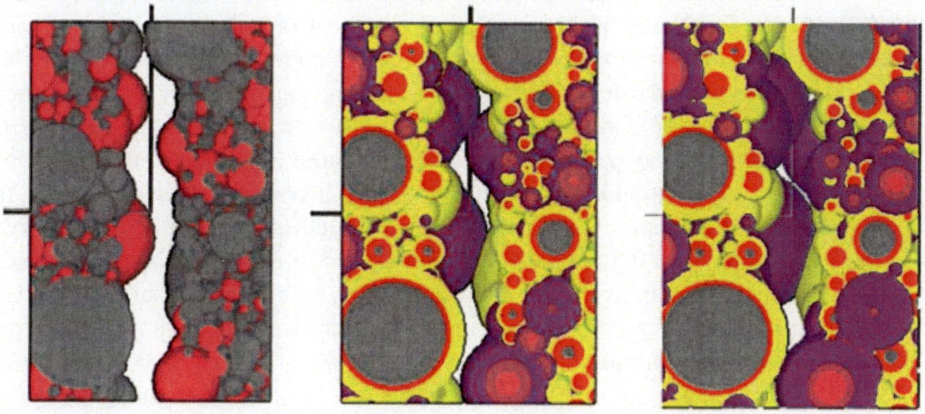

Fig. 10 Microstructure with 30% replacement of cement particles by DEP with a crack width of 30 μm. (left) The initial state of the particle structure, (centre) the initiated crack with a hydration of the initial particles up to 40 days and DEPs starting hydration after 28 days and (right) similar with a double expansion potential of the DEPs (reproduced from [94], with permission)

Huang and Ye [88] developed a model for the simulation of a self-healing system that employed encapsulated water to promote self-healing through further hydration. In their simulation, they consider a 10 μm wide micro-crack passing through a microcapsule in the presence of some unhydrated cement particles. The authors formulated a reactive transport model that accounted for the loss of moisture from the crack into the cement matrix, the diffusion of chemical species throughout the crack domain, and the chemical reactions governing the self-healing process. The initial distribution of unhydrated cement particles was calculated using Hymostruc3D, whilst the hydration reactions were simulated using a thermodynamic model. The thermodynamic model consisted of the charge and mass balance equations, coupled to the equilibrium equations that describe the chemical reactions considered. The simulation considered further hydration forming both CS and CSH. A closed system was assumed, such that the entry of carbon dioxide was blocked and calcium carbonate precipitation could be neglected. The model also accounted for the effect of the precipitation of material in the crack on the transport properties. The results of the simulations indicated that encapsulated water formed an effective means to promote self-healing but that the precipitation of material in the crack can inhibit the transport of ions, reducing the potential for further healing.

Zemskov et al. [67] took an interesting approach to the simulation of bacterial self-healing of cracks. Their model considered a crack that crossed through an embedded microcapsule that contains calcium lactate, sodium glutamate and monopotassium phosphate. The first of these compounds is the reactant that is converted to calcium carbonate to fill the crack. The latter two are nutrients supplied for the bacteria. The crack is assumed to be instantaneously filled with water, and as such, transport of the chemical species is governed by a diffusion equation. The focus of the work is on the moving interfaces involved during the chemical reaction. The interfaces correspond to the front of precipitated calcium carbonate within the crack (that moves outwards from the crack face with further precipitation), the front of the capsule contents (that moves inwards with further consumption of calcium lactate) and the interface between the precipitated calcium carbonate and the capsule boundary. The movement of the interfaces is governed by the reactive transport problem and is tracked using the level-set method. The growth of bacteria is accounted for using a Monod model. The results of a simulation based on the healing of a crack by a single capsule for different crack widths and capsule radii are presented. As an example, the model predicts that capsules of 0.5 mm radius with about 1.5 mm distance between capsule centres can successfully heal a 0.1 mm crack. The authors note that the effect of neighbouring capsules can be inferred and that generalisation of the model to three dimensions is straightforward.

A model for the simulation of bio-mineralisation in SHCMs was presented by Algaifi et al. [69]. The model considered a fictitious crack that passed through an embedded capsule and was instantaneously saturated. Bacteria cells, nutrients and calcium ions were distributed throughout the crack, with bacteria cell concentrations varying from an optimum concentration of 10×10^8 cell/cm^3 at the crack mouth where more oxygen can be found, to 10×10^5 cell/cm^3 in the deepest part of the crack. A schematic of the conceptual model can be seen in Fig. 11. The transport of urea

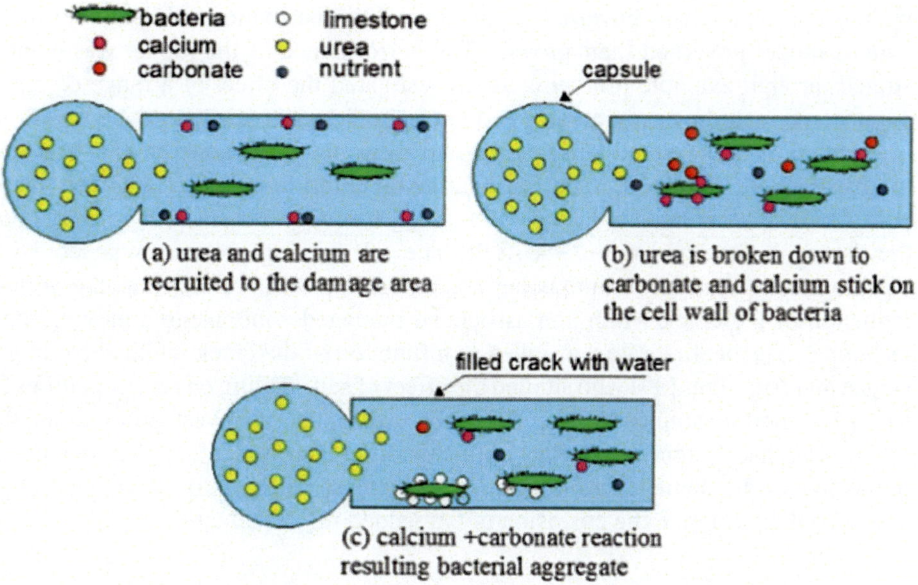

Fig. 11 Schematic of the conceptual model (reproduced from [69], with permission)

from the embedded capsule was described using Fick's law, whilst the hydrolysis and calcium carbonate precipitation reactions were modelled using first-order kinetics. The model results compared reasonably with experimental data, predicting complete filling of the crack with calcium carbonate at 60 days, compared to the experimentally observed 70 days.

The influence of self-healing on the chloride transport in cementitious materials was investigated in a study by Jiang et al. [95]. The cement hydration and microstructure development were simulated for ranges of different mixes for 28 days using CEMHYD3D [96], before an artificial crack was introduced. Different cement mixes were considered, along with a range of crack widths. To simulate the self-healing, the hydration and microstructure development was further simulated following the introduction of the crack. The chloride diffusion coefficient of the simulated microstructures at different healing times was calculated using a random walk model. It was found that the chloride diffusion coefficient had only a small dependence on the mixture proportions, but that there was a large dependence on crack width, and subsequently the degree of healing.

A mesoscale lattice Boltzmann model for the simulation of self-healing through further hydration was presented in Chen and Ye [97]. The simulation domain was obtained using analysis on an image obtained from an experiment. Following the approach of Chitez and Jefferson [60], the unhydrated cement was considered as a single solute, whilst precipitated phases were not considered individually, but instead as high-density and low-density hydration products. The nucleation of the hydration products was governed by a probability distribution, whilst a thermodynamics model governed their growth. The performance of the model was demonstrated through example problems that investigated the effect of a range of model parameters, including growth rates and nucleation constants.

Xu et al. [98] presented a five-phase mesoscale model of concrete that was used to investigate the effect of crack self-healing on chloride diffusion. The five phases considered corresponded to the coarse aggregate, mortar, interfacial transition zone, the damaged zone and the crack itself. The self-healing process considered was of autogenous healing of an existing crack. Healing was simulated as the gradual reduction of the crack width and associated damaged zone, using a moving mesh technique. The healing rate was given as a function of the crack width according to Gagné and Argouges [99] who studied the effect of self-healing on crack openings for a range of mortar samples using airflow measurements. The investigation found that self-healing had a significant effect on the chloride distribution and that an increase in the size of the damaged zone leads to an increase in the chloride concentration, which could accelerate the corrosion of embedded reinforcement.

3.4 *Optimisation of Embedded Vascular Networks*

The optimum design of an embedded vascular network is a complex problem that depends upon a number of factors. For instance, the network should be designed such that the probability of a crack crossing a channel, triggering healing, is high. In addition to this, the efficiency of the flow through the network should be as high as possible, whilst blockages, and weakening of the cementitious matrix through the introduction of too many channels, are to be avoided.

Bejan et al. [100] considered the optimisation of a network of channels for a self-healing composite material. The investigation considered a fixed volume of channels that must be configured in a grid structure. The optimised quantity was chosen as the permeability of the system, as this leads to a faster crack filling. Square, triangular and hexagonal grids comprising channels of equal diameter, and a square grid comprising channels of two different diameters were considered. It was found that the best grid structure was triangular, whilst a grid with two optimised channel sizes performed better than those with one. Aragón et al. [101] considered the optimisation of a three-dimensional microvascular network. The model coupled a multi-objective constrained genetic algorithm [102] with a finite element model to predict the effect of the microvascular network on the temperature distribution. The objectives were to minimise the void fraction left by the network, pressure drop across the network and the maximum temperature in the component, whilst a constraint was set on the connectivity of the network. The resulting model was applied to two example problems concerning a periodic microvascular cell and a microvascular fin with localised heating. An example of the optimised networks produced can be seen in Fig. 12. Whilst the model was not developed with SHCMs in mind, it is clear that the approach could prove useful in the design of vascular self-healing systems with the appropriate definition of objective functions and constraints.

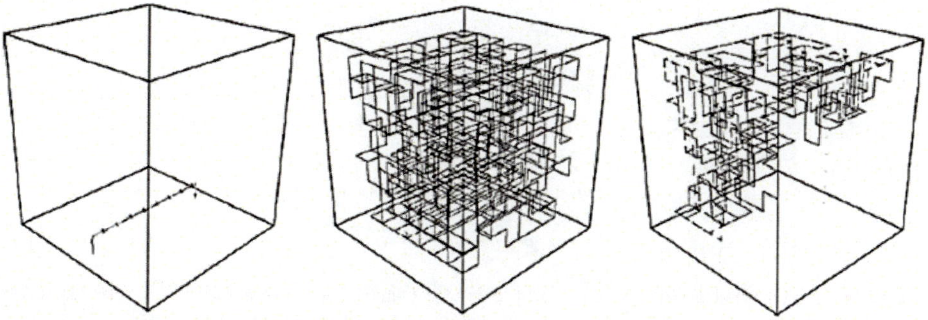

Fig. 12 Optimised microvascular networks showing the optimal solution for (left) void fraction, (centre) pressure drop and (right) temperature (reproduced from [101], with permission)

4 Coupled Models for the Simulation of SHCMs

Now that the simulation of the different physical processes that govern the response of SHCMs has been discussed, it was considered useful to review fully coupled models for the simulation of the mechanical, transport and healing behaviour. There has been little work to date on such coupled models developed for SHCMs, though this state of affairs is now changing and such models have also been developed with other material systems in mind (see for example [103, 104].

A combined experimental and numerical investigation into the autogenous healing of small cracks (5–20 μm) was presented in Hillouin et al. [105]. The numerical investigation employed a modified version of CemPy (CEMHYD3D) called CemPP to simulate the microstructure development. The hydration process was modelled using cycles of dissolution, diffusion and reactions. The healing process concerned the further hydration of unreacted clinker that led to a precipitation of hydration products on the face of a simulated crack. The simulated microstructure served as the input for the mechanical model, Cast3M, that was utilised to study the mechanical regains due to the healing. The time-varying mechanical properties were calculated by successively calculating Young's moduli of the microstructure in directions normal and perpendicular to the crack. Assuming constant mechanical properties of healing products and that the healing is symmetric with respect to the crack, the authors found that the regains of mechanical strength and stiffness are directly related to the filling fraction at the centre of the crack, as shown in Fig. 13.

Di Luzio et al. [106] presented a coupled model for self-healing cementitious composites. The model coupled the SMM (solidification–microprestress–microplane model M4) model for concrete, which the authors extended to incorporate self-healing effects, to a hygro-thermo-chemical model to describe the transport [107]. The driving force for the healing was described using a reaction affinity that considered delayed cement hydration, which is the main cause of self-healing in young concrete, along with the effects of additives. The reaction rate considered the effects of a number of factors, including temperature, hydration degree, relative humidity and crack width. The model considered both the effects of cracking on the moisture

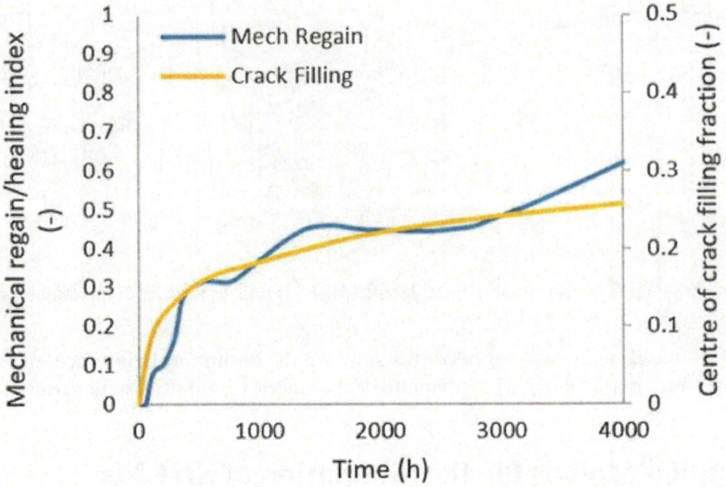

Fig. 13 Model predictions of mechanical regains and centre of crack filling fraction (after [105])

transport, and the regain in mechanical properties as a result of self-healing. The models were satisfactorily validated against the experimental results of Ferrara et al. [108].

Sanz-Herrera et al. [109] presented a coupled mechanochemical-diffusive model for self-healing materials. Their model coupled the chemical-diffusive chemical model presented in [68], with a CDHM model that employs a rate-dependant damage evolution based on that presented by Darabi et al. [13]. The self-healing mechanism concerned the reaction of a mobile chemical species, diffusing in from the environment, with a static substance present in the matrix material to precipitate a healing product. The degree of healing was given as the volume fraction of precipitated materials in the crack. The performance of the model was demonstrated through an example problem concerning the three-point bending of a beam subject to cycles of loading/unloading and healing periods.

Romero Rodríguez et al. [110] developed a mesoscale lattice model for the simulation of crack-sealing using super-absorbent polymers (SAPs). The moisture transport was simulated using the diffusive form of Richard's equation, which has the advantage that the diffusivity may be estimated from the sorptivity. The effect of cracking on the diffusivity was taken into account using a relative permeability that was calculated using Poiseuille's law for flow in a planar crack. The cracking was simulated using a lattice fracture model, with the resulting fractured elements being assigned a crack width of 200 μm, based on experimental measurements [111]. The swelling of SAPs was simulated using the swelling law found in [112], which is based on the swelling capacity of an individual SAP. The model was calibrated and validated using experimental data, before a parametric study was undertaken to determine the effect of absorption capacity, dosage and crack widths on the predicted response. The results showed that increasing the swelling ratio of SAPs will lead to more efficient crack-sealing than increasing the SAP dosage.

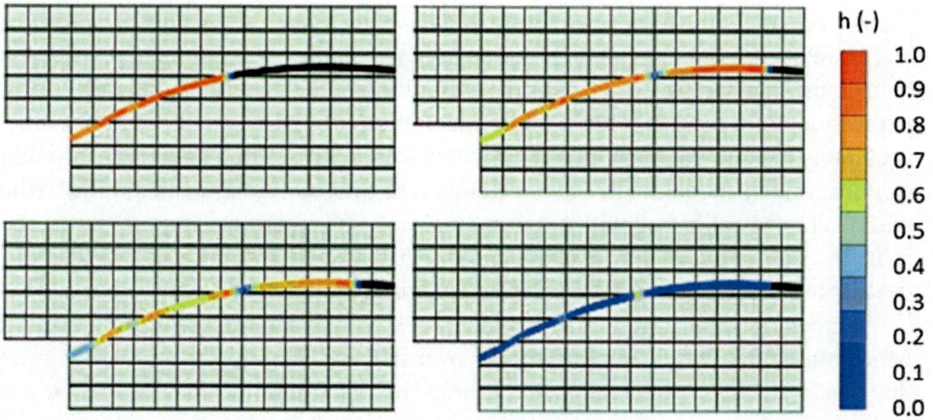

Fig. 14 Contour plots of the degree of healing (increasing time from top left to bottom right, the black line represents empty part of the crack) (reproduced from [24])

Freeman and Jefferson [113] (see also [24]) presented a coupled model for the simulation of SHCMs with embedded healing systems containing an autonomic healing agent. The transport processes considered the discrete crack flow, described using the Navier–Stokes equations, coupled to the continuum matrix flow, described using Richard's equation. The effect of dynamic contact angle, non-uniformity of the cracks and the effect of healing on the rheological properties were taken into account. The cracks were represented using an embedded strong-discontinuity element, whilst a damage-healing model that allows for simultaneous cracking and healing, along with multiple re-cracking and re-healing events, was employed to describe the constitutive behaviour. The healing was simulated using a kinetic reaction model based on the propagation of curing fronts emanating from the crack faces. The model was calibrated and subsequently validated against data from a linked experimental programme [114, 115]. The comparison showed that the model was able to reproduce the characteristic mechanical and flow behaviour for a range of crack configurations and load paths. An example of the predicted healing response for the loading of an L-shaped specimen can be seen in Fig. 14, that shows the crack path and predicted degree of healing.

5 Discussion

Significant progress has been made on the development of numerical models for simulating SHCMs. Nevertheless, there are still a number of challenges that need to be addressed.

One such challenge is the validation of models using experimental data. If numerical models are to be used in the development of SHCMs, as well as in the design and analysis of structures formed from SHCMs, then it is important to quantify their predictive capability. Where validations have been made, the results are promising and show the potential of models for accurately capturing experimentally observed behaviour. A related issue is that of the calibration of the material parameters utilised in the models. A comprehensive numerical model for SHCMs will need to be able to accurately capture the many interacting physical processes that govern their material response. Such models will have a number of different components, all of which will need to be calibrated with experimental data. It is therefore important that there is interaction between the researchers developing and testing SHCMs and those developing numerical models to predict and analyse their behaviour.

Another key challenge lies in the fact that the physical processes that govern the behaviour of SHCMs, frequently occur over multiple length scales. An example of this can be found in microcapsule-based SHCMs, where the material response at the macroscale is governed by processes that occur at the microscale. For instance, the rupture of microcapsules and changes to the microstructure with damage and healing affect not only the strength and stiffness observed at the macroscale, but also the transport properties. In addition to this, the microstructure and the material properties of SHCMs show a significant variability that has only been considered in a few investigations. The issue of multiple length scales will be exacerbated if numerical models are to be used in the design and analysis of structures formed from SHCMs, where simulations will be required at the structural scale.

6 Concluding Remarks

There is now a significant body of research on the numerical simulation of SHCMs. The research that has been undertaken to date has shown that numerical models are capable of accurately capturing the behaviour of the many different individual aspects of SHCMs. There are, however, few fully coupled models that have been developed to date to simulate the many physical processes that govern the behaviour of a particular SHCM. Furthermore, the validation of many of the models presented is limited. Overall, the authors consider that major advances have been made, but that the journey towards the development of comprehensive models for the simulation of SHCMs is in its early stages.

References

1. Lemaitre J (1985) A continuum damage mechanics model for ductile fracture. J Eng Mater Technol Trans ASME 107(1):83–89
2. Mazars J (1986) A description of micro- and macroscale damage of concrete structures. Eng Fract Mech 25(5–6):729–737
3. Simo JC, Ju JW (1987) Strain- and stress-based continuum damage models-I. Formulation. Int J Solids Struct 23(7):821–840
4. Lemaitre J, Desmorat R (2005) Engineering damage mechanics. Springer

5. Gawin D, Pesavento F, Schrefler BA (2003) Modelling of hygro-thermal behaviour of concrete at high temperature with thermo-chemical and mechanical material degradation. Comput Methods Appl Mech Eng 192(13–14):1731–1771
6. Comi C, Perego U (2001) Fracture energy based bi-dissipative damage model for concrete. Int J Solids Struct 38:6427–6454
7. Abu Al-Rub RK, Darabi MK (2012) A thermodynamic framework for constitutive modeling of time- and rate-dependent materials. Part I: theory. Int J Plast 34:61–92
8. Barbero EJ, Greco F, Lonetti P (2005) Continuum damage-healing mechanics with application to self-healing composites. Int J Damage Mech 14:51–81
9. Granger S, Pijaudier-Cabot G, Loukili A (2007) Mechanical behaviour of self-healed ultra high performance concrete: from experimental evidence to modeling. In: Proceedings of the 6th international conference on fracture mechanics of concrete and concrete structures, vol 3, pp 1827–1834
10. Mergheim J, Steinmann P (2013) Phenomenological modelling of self-healing polymers based on integrated healing agents. Comput Mech 52(3):681–692
11. Abu Al-Rub RK, Darabi MK, Little DN, Masad EA (2010) A micro-damage healing model that improves prediction of fatigue life in asphalt mixes. Int J Eng Sci 48:966–990
12. Darabi MK, Abu Al-Rub RK, Masad EA, Little DN (2012) A thermodynamic framework for constitutive modeling of time- and rate-dependent materials. Part II: numerical aspects and application to asphalt concrete. Int J Plast 35:67–99
13. Darabi MK, Abu Al-Rub RK, Little DN (2012) A continuum damage mechanics framework for modelling micro-damage healing. Int J Solids Struct 49:492–513
14. Esgandani GA, El-Zein A (2020) Thermodynamic based model for coupled elastoplastic damage-healing behaviour of unsaturated geomaterials. Mech Mater 145:103395
15. Oucif C, Voyiadjis GZ, Rabczuk T (2018) Modeling of damage-healing and nonlinear self-healing concrete behaviour: application to coupled and uncoupled self-healing mechanisms. Theoret Appl Fract Mech 96:216–230
16. Oucif C, Voyiadjis GZ, Kattan PI, Rabczuk T (2019) Investigation of the super healing theory in continuum damage and healing mechanics. Int J Damage Mech 28(6):896–917
17. Voyiadjis GZ, Kattan PI (2014) Healing and super healing in continuum damage mechanics. Int J Damage Mech 23(2):245–260
18. Joseph C, Jefferson A, Isaacs B, Lark R, Gardner D (2010) Experimental investigation of adhesive-based self-healing of cementitious materials. Mag Concr Res 62(11):831–843
19. Voyiadjis GZ, Shojaei A, Li G (2011) A thermodynamic consistent damage and healing model for self healing materials. Int J Plast 27(7):1025–1044
20. Voyiadjis GZ, Shojaei A, Li G, Kattan PI (2012) A theory of anisotropic healing and damage mechanics of materials. Proc R Soc A Math Phys Eng Sci 468(2137):163–183
21. Voyiadjis GZ, Shojaei A, Li G (2012) Continuum damage-healing mechanics with introduction to new healing variables. Int J Damage Mech 21:391–414
22. Voyiadjis GZ, Kattan PI (2017) Mechanics of damage, healing, damageability, and integrity of materials: a conceptual framework. Int J Damage Mech 26(1):50–103
23. Voyiadjis GZ, Kattan PI (2018) Decomposition of healing tensor: in continuum damage and healing mechanics. Int J Damage Mech 27(7):1020–1057
24. Freeman BL, Bonilla-Villalba P, Mihai IC, Alnaas WF, Jefferson AD (2020) A specialised finite element for simulating self-healing quasi-brittle materials. Adv Model Simul Eng Sci 7:32
25. Hillerborg A, Modeer M, Petersson PE (1976) Analysis of crack formation and crack growth in concrete by means of fracture mechanics and finite elements. Cem Concr Res 6(6):773–781
26. Schimmel EC, Remmers J (2006) Development of a constitutive model for self-healing materials. Delft Aerospace Computational Science, Report DACS-06-003
27. Remmers J, de Borst R (2008) Numerical modelling of self-healing mechanisms. In: Self healing materials. Springer, pp 365–380
28. Maiti S, Shankar C, Geubelle PH, Kieffer J (2006) Continuum and molecular-level modeling of fatigue crack retardation in self-healing polymers. J Eng Mater Technol 128:595–602

29. Alsheghri AA, Abu Al-Rub RK (2015) Thermodynamic-based cohesive zone healing model for self-healing materials. Mech Res Commun 70:102–113
30. Alsheghri AA, Abu Al-Rub RK (2016) Finite element implementation and application of a cohesive zone damage-healing model for self-healing materials. Eng Fract Mech 163:1–22
31. Jud K, Kausch HH (1979) Load transfer through chain molecules after interpenetration at interfaces. Polym Bull 1(10):697–707
32. Caggiano A, Etse G, Ferrara L, Krelani V (2017) Zero-thickness interface constitutive theory for concrete self-healing effects. Comput Struct 186:22–34
33. Zhang Y, Zhuang X (2018) A softening-healing law for self-healing quasi-brittle materials: analysing with strong discontinuity embedded approach. Eng Fract Mech 192:290–306
34. Ponnusami SA, Krishnasamy J, turteltaub S, van der Zwaag S (2018) A cohesive-zone crack healing model for self-healing materials. Int J Solids Struct 134:249–263
35. Xin A, Du H, Yu K, Wang Q (2020) Mechanics of bacteria-assisted extrinsic healing. J Mech Phys Solids 39:013983
36. Li W, Jiang Z, Yang Z, Yu H (2016) Effective mechanical properties of self-healing cement matrices with microcapsules. Mater Des 95:422–430
37. Eshelby JD (1957) The determination of the elastic field of an ellipsoidal inclusion and related problems. Proc R Soc A Math Phys Eng Sci 241:376–396
38. Ms Q, Zhuang X, Rabczuk T (2015) Computational model generation and RVE design of self-healing concrete. Front Struct Civ Eng 9(4):383–396
39. Davies R, Jefferson AD (2017) Micromechanical modelling of self-healing cementitious materials. Int J Solids Struct 113–114:180–191
40. Herbst O, Luding S (2008) Modeling particulate self-healing materials and application to uni-axial compression. Int J Fract 154:87–103
41. Zhou S, Zhu H, Ju JW, Yan Z, Chen Q (2017) Modeling microcapsule-enabled self-healing cementitious composite materials using discrete element method. Int J Damage Mech 26(2):340–357
42. Gilabert FA, Garoz D, Van Paepegem W (2017) Macro-and micro-modeling of crack propagation in encapsulation-based self-healing materials: application of XFEM and cohesive surface techniques. Mater Des 130:459–478
43. Šavija B, Feiteira J, Araújo M, Chatrabhuti S, Raquez J-M, Van Tittelboom K, Gruyaert E, De Belie N, Schlangen E (2017) Simulation-aided design of tubular polymeric capsules for self-healing concrete. Materials 10(1):10
44. Xue C, Li W, Wang K, Sheng D, Shah SP (2020) Novel experimental and numerical investigations on bonding behaviour of crack interface in smart self-healing concrete. Smart Mater Struct 29:085004
45. Mauludin LM, Budiman BA, Santosa SP, Zhuang X, Rabczuk T (2020) Numerical modeling of microcrack behaviour in encapsulation-based self-healing concrete under uniaxial tension. J Mech Sci Technol 34(5):1847–1853
46. Mauludin LM, Oucif C (2018) Interaction between matrix crack and circular capsule under uniaxial tension in encapsulation-based self-healing concrete. Underground Space 3:181–189
47. Mauludin LM, Oucif C (2019) The effects of interfacial strength on fractured microcapsule. Front Struct Civil Eng 13(2):353–363
48. Krishnasamy J, Ponnusami SA, Turteltaub S, van der Zwaag S (2018) Modelling the fracture behaviour of thermal barrier coatings containing healing particles. Mater Des 157:75–86
49. Ponnusami SA, Krishnasamy J, Turteltaub S, van der Zwaag S (2019) A micromechanical fracture analysis to investigate the effect of healing particles on the overall mechanical response of a self-healing particulate composite. Fatigue Fract Eng Struct Mater 42:533–545
50. Zemskov SV, Jonkers HM, Vermolen FJ (2011) Two analytical models for the probability characteristics of a crack hitting encapsulated particles: application to self-healing materials. Comput Mater Sci 50:3323–3333
51. Huang H, Ye G (2016) Numerical studies of the effects of water capsules on self-healing efficiency and mechanical properties in cementitious materials. Adv Mater Sci Eng 8271214

52. Wang M, Hu X, Zhao Y (2020) Probabilistic analysis models to determine capsule dosage for healing of cracks in concrete. Adv Struct Eng 1–13
53. Hazelwood T, Jefferson AD, Lark R, Gardner D (2015) Numerical simulation of the long-term behaviour of a self-healing concrete beam vs standard reinforced concrete. Eng Struct 102:176–188
54. Dunn SC, Jefferson AD, Lark R, Isaacs B (2011) Shrinkage behaviour of poly(ethylene terephthalate) for a new cementitious-shrinkable polymer material system. J Appl Polym Sci 120(5):2516–2526
55. Hazelwood T, Jefferson AD, Lark R, Gardner D (2014) Long-term stress relaxation behaviour of predrawn poly(ethylene terephthalate). J Appl Polym Sci 131(23)
56. Balzano B, Sweeney J, Thompson G, Jefferson AD (2021) Enhanced concrete crack closure with hybrid shape memory polymer tendons. Eng Struct 226, 111330. https://doi.org/10.17035/d.2020.00874063774399-4410
57. Wang Y, Ding W, Fang G, Liu Y, Xing F, Dong B (2016) Feasibility study on corrosion protection of steel bar in a self-immunity system based on increasing OH-content. Constr Build Mater 125:742–748
58. Lewis RW, Schrefler BA (1998) The finite element method in the static and dynamic deformation and consolidation of porous media, 2nd edn. Wiley Inc., Great Britian
59. Koniorczyk M, Gawin D (2008) Heat and moisture transport in porous building materials containing salt. J Building Phys 31(4):279–300
60. Chitez AS, Jefferson AD (2016) A coupled thermo-hygro-chemical model for characterising autogenous healing in ordinary cementitious materials. Cem Concr Res 88:184–197
61. Jennings HM (2008) Refinements to colloid model of C-S-H in cement: CM-II. Cem Concr Res 38:275–289
62. Chitez AS, Jefferson A (2015) Porosity development in a thermo-hygral finite element model for cementitious materials. Cem Concr Res 78:216–233
63. Papadakis VG, Vayenas CG, Fardis MN (1989) A reaction-engineering approach to the problem of concrete carbonation. AIChE J 35:1639–1650
64. Saetta AV, Schrefler BA, Vitaliani RV (1993) The carbonation of concrete and the mechanism of moisture, heat and carbon dioxide flow through porous materials. Cem Concr Res 23:761–772
65. Steffens A, Dinkler D, Ahrens H (2002) Modeling carbonation for corrosion risk prediction of concrete structures. Cem Concr Res 32:935–941
66. Peter MA, Muntean A, Meier SA, Böhm M (2008) Competition of several carbonation reactions in concrete: a parametric study. Cem Concr Res 38:1385–1393
67. Zemskov SV, Jonkers HM, Vermolen FJ (2014) A mathematical model for bacterial self-healing of cracks in concrete. J Intell Mater Syst Struct 25(1):4–12
68. Aliko-Benítez A, Doblare M, Sanz-Herrera JA (2015) Chemical-diffusive modelling of the self-healing behaviour in concrete. Int J Solids Struct 69–70:392–402
69. Algaifi HA, Bakar AB, Sam ARM, Abidin ARZ, Shahir S, Al-Towayti WAH (2018) Numerical modeling for crack self-healing concrete by microbial calcium carbonate. Constr Build Mater 189:816–824
70. Javierre E, Gaspar FJ, Rodrigo C (2019) Modelling the carbonation reactions in self-healing concrete. In: Proceedings of the 10th international conference on fracture mechanics of concrete and concrete structures, p 235637
71. Roels S, Vandersteen K, Carmeliet J (2003) Measuring and simulating moisture uptake in a fractured porous medium. Adv Water Resour 26(3):237–246
72. Roels S, Moonen P, De Proft K, Carmeliet J (2006) A coupled discrete-continuum approach to simulate moisture effects on damage processes in porous materials. Comput Methods Appl Mech Eng 195(52):7139–7153
73. Carmeliet J, Delerue JF, Vandersteen K, Roels S (2004) Three-dimensional liquid transport in concrete cracks. Int J Numer Anal Meth Geomech 8(7–8):671–687
74. Hall J, Qamar IPS, Rendall TCS, Trask RS (2015) A computational model for the flow of resin in self-healing composites. Smart Mater Struct 24:037002

75. Gardner D, Jefferson AD, Hoffman A (2012) Investigation of the capillary flow in discrete cracks in cementitious materials. Cem Concr Res 42(7):972–981
76. Gardner D, Jefferson AD, Hoffman A, Lark R (2014) Simulation of the capillary flow of an autonomic healing agent in discrete cracks in cementitious materials. Cem Concr Res 58:35–44
77. Gardner D, Herbert D, Jayaprakash M, Jefferson AD, Paul A (2017) Capillary flow characteristics of an autogenic and autonomic healing agent for self-healing concrete. J Mater Civil Eng 29(11):4017228
78. Gilabert FA, Van Tittelboom K, Van Stappen J, Cnuddle V, De Belie N, Van Paepegem W (2017) Integral procedure to assess crack filling and mechanical contribution of polymer-based healing-agent in encapsulation-based self-healing concrete. Cement Concr Compos 77:68–80
79. Jasak H (2009) OpenFOAM: open source CFD in research and industry. Int J Naval Arch Ocean Eng 1(2):89–94
80. Ranaivomanana H, Benkemoun N (2017) Numerical modelling of the healing process induced by carbonation of a single crack in concrete structures: theoretical formulation and embedded finite element method implementation. Finite Elem Anal Des 132:42–51
81. Iyer J, Walsh SDC, Hao Y, Carroll SA (2018) Assesment of two-phase flow on the chemical alteration and sealing of leakage pathways in cemented wellbores. Int J Greenhouse Gas Control 69:72–80
82. Walsh SDC, Mason HE, Frane WLD, Carroll SA (2014) Experimental calibration of a numerical model describing the alteration of cement/caprock interfaces by carbonated brine. Int J Greenhouse Gas Control 22:176–188
83. Walsh SDC, Mason HE, Frane WLD, Carroll SA (2014) Mechanical and hydraulic coupling in cement-caprock interfaces exposed to carbonated brine. Int J Greenhouse Gas Control 22:176–188
84. Iyer J, Walsh SDC, Hao Y, Carroll SA (2017) Incorporating reaction-rate dependence in reaction-front models of wellbore-cement/carbonated-brine systems. Int J Greenhouse Gas Control 59:160–171
85. van Breugel K (1995) Numerical simulation of hydration and microstructural development in hardening cement based materials. Cem Concr Res 25:319–331
86. Ter Heide N (2005) Crack healing in hydrating concrete. MSc thesis, Delft University of Technology
87. Ye G, van Breugel K (2007) Potential use of hymostruc cement hydration model for self-healing of microcracks in cementitious materials. In: Proceedings of the first international conference on self healing materials. Noordwijk aan Zee, The Netherlands
88. Huang H, Ye G (2012) Simulation of self-healing by further hydration in cementitious materials. Cement Concr Compos 34(4):460–467
89. Koenders EAB (2012) Modelling the self-healing potential of dissoluble encapsulated cement systems, Final report IOP project SHM08707
90. Koenders EAB (1997) Simulation of volume changes in hardening cement-based materials. PhD thesis, Delft University Press
91. Lokhorst SJ (1999) Deformational behavior of concrete influenced by hydration related changes of the microstructure. Research report, Delft University of Technology
92. Koenders EAB, Ukrainczyk N, Caggiano A (2018) Modelling the self-healing potential of dissoluble encapsulated cement. In: Proceedings of the 6th European conference on computational mechanics—7th European conference on computational fluid dynamics, pp 721–733
93. Yang S, Caggiano A, Yi M, Ukrainczyk N, Koenders EAB (2019) Modelling autogenous self-healing with dissoluble encapsulated particles using a phase field approach. In: Proceedings de congresos organizados por la Asociación Argentina de Mecánica Computacional AMCA volume XXXVII, pp 1457–1467
94. Jefferson AD, Javierre E, Freeman BL, Zaoui A, Koenders EAB, Ferrara L (2018) Research progress on numerical models for self-healing cementitious materials. Adv Mater Interfaces 5:1701378

95. Jiang J, Zheng X, Wu S, Liu Z, Zheng Q (2019) Nondestructive experimental characterization and numerical simulation on self-healing and chloride ion transport in cracked ultra-high performance concrete. Constr Build Mater 198:696–709
96. Bentz DP (2005) CEMHYD3D: a three-dimensional cement hydration and microstructure development modeling package. NIST Interagency Internal Report (NISTIR) 7232
97. Chen J, Ye G (2019) A lattice Boltzmann single component model for simulation of the autogenous self-healing caused by further hydration in cementitious materials at mesoscale. Cement Concrete Res 123:105782
98. Xu J, Peng C, Wan L, Wu Q, She W (2020) Effect of crack self-healing on concrete diffusivity: mesoscale dynamics simulation study. J Mater Civ Eng 32(6):04020149
99. Gagné R, Argouges M (2012) A study on the natural self-healing of mortars using air-flow measurements. Mater Struct 45(11):1625–1638
100. Bejan A, Lorente S, Wang KM (2006) Networks of channels for self-healing composite materials. J Appl Phys 100:033528
101. Aragón AM, Saksena R, Kozola BD, Ph, Geubelle, Christensen KT, White SR (2013) Multi-physics optimization of three-dimensional microvascular polymeric components. J Comput Phys 233:132–147
102. Aragón AM, Wayer JK, Geubelle PH, Goldberg DE, White SR (2008) Design of microvascular flow networks using multi-objective genetic algorithms. Comput Methods Appl Mech Eng 197:4399–4410
103. Bluhm J, Specht S, Schröder J (2015) Modelling of self-healing effects in polymeric composites. Arch Appl Mech 85:1469–1481
104. Jia SP, Zhang LW, Wu BS, Yu HD, Shu JX (2018) A coupled hydro-mechanical creep damage model for clayey rock and its application to nuclear waste repository. Tunn Undergr Space Technol 74:230–246
105. Hilloulin B, Hilloulin D, Grondin F, Loukili A, De Belie N (2016) Mechanical regains due to self-healing in cementitious materials: experimental measurements and micro-mechanical model. Cem Concr Res 80:21–32
106. Di Luzio G, Ferrara L, Krelani V (2018) Numerical modeling of mechanical regain due to self-healing in cement based composites. Cement Concr Compos 86:190–205
107. Di Luzio G, Cusatis G (2009) Hygro-thermo-chemical modeling of high performance concrete. I: theory. Cement Concr Compos 31:301–308
108. Ferrara L, Krelani B, Carsana M (2014) A fracture testing based approach to assess crack healing of concrete with and without crystalline admixtures, 68:515–531
109. Sanz-Herrera JA, Aliko-Benítez A, Fadrique-Contreras AM (2019) Numerical investigation of the coupled mechanical behaviour of self-healing materials under cyclic loading. Int J Solids Struct 160:232–246
110. Romero Rodríquez C, Chaves Figueiredo S, Deprez M, Snoeck D, Schlangen E, Šavija B (2019) Numerical investigation of crack self-sealing in cement-based composites with superabsorbent polymers. Cement Concrete Comp 104:103395
111. Snoeck D (2015) Self-healing and microstructure of cementitious materials with microfibers and superabsorbent polymers. PhD thesis, Ghent University
112. Esteves LP (2011) Superabsorbent polymers: on their interaction with water and pore fluid. Cement Concr Compos 33(7):717–724
113. Freeman BL, Jefferson AD (2020) The simulation of transport processes in cementitious materials with embedded healing systems. Int J Numer Anal Meth Geomech 44:293–326
114. Selvarajoo T, Davies RE, Gardner D, Freeman BL, Jefferson AD (2020) Characterisation of a vascular self-healing cementitious material system: flow and curing properties. Construct Build Mater 245:118332
115. Selvarajoo T, Davies RE, Freeman BL, Jefferson AD (2020) Mechanical response of a vascular self-healing cementitious material system under varying load conditions. Construct Build Mater 254:119245

Modeling of Self-healing Process in Bituminous Materials: Experimental and Numerical Models

Guoqiang Sun, Daquan Sun, Mingjun Hu, Alvaro Guarin, and Jose Norambuena-Contreras

Abstract This chapter presents the relevant research details and progress of the current three types of self-healing models in bituminous materials based on the relevant literature, i.e. physical–chemical-based experimental model, physical–chemical-based numerical model and mechanical-based self-healing modelmechanical-based self-healing model. Moreover, the development and future modeling outlook of the self-healing process in bituminous materials are discussed.

1 Introduction

The self-healing phenomenon in materials was a very complicated physical–chemical process. Most investigators applied mechanical experiments to comprehend the

G. Sun
Beijing Key Laboratory of Traffic Engineering, Beijing University of Technology, Beijing, P.R. China
e-mail: gqsun@bjut.edu.cn

D. Sun (✉) · M. Hu
Key Laboratory of Road and Traffic Engineering of Ministry of Education, Tongji University, Shanghai, P.R. China
e-mail: dqsun@tongji.edu.cn

M. Hu
e-mail: humj@tongji.edu.cn

A. Guarin
Department of Civil and Architectural Engineering, KTH Royal Institute of Technology, Stockholm, Sweden
e-mail: alvaro.guarin@abe.kth.sehumj

J. Norambuena-Contreras
Department of Civil and Environmental Engineering, LabMAT, University of Bío-Bío, Concepción, Chile
e-mail: jnorambuena@ubiobio.cl

© Springer Nature Switzerland AG 2022

A. Kanellopoulos and J. Norambuena-Contreras (eds.), *Self-Healing Construction Materials*, Engineering Materials and Processes,
https://doi.org/10.1007/978-3-030-86880-2_7

self-healing behavior. Nevertheless, the results of self-healing ability obtained were inconsistent and depended on the test methods, test conditions, and evaluation standard. Therefore, a large number of scholars underlined the significance of comprehending the self-healing mechanism through modeling. In this context, three kinds of self-healing models were summarized based on the relevant literature.

The first class of self-healing model was called as *physical–chemical-based experimental model*, which was commonly used to explore the self-healing mechanism based on some empirical phenomena, including the inter-molecular diffusion self-healing model, the surface energy self-healing model, and the capillary flow self-healing model. The inter-molecular diffusion self-healing mechanism in bituminous materials was developed from the polymer self-healing behavior theory in accordance with polymer chain dynamics [1–3]. The self-healing process in bituminous materials could be divided into five microscopic phases in the light of the inter-molecular diffusion theory [4–6]. The surface energy self-healing model was developed by Lytton et al. [7] and Schapery [8]. They established the relationship between the self-healing level and the surface energy of bitumen and put forward a self-healing rate function [9, 10]. Regarding a new open macro-crack in bituminous materials, the inter-molecular diffusion would not happen spontaneously because of the large width between the cracking faces. Nonetheless, the self-healing process could be continuously occurring, and then, the visible crack gradually disappeared when sufficient rest time and heating energy were introduced. Consequently, the capillary flow self-healing model was established to describe the meso-crack self-healing or even macro-crack self-healing potential by the modification for Lucas–Washburn equation [11–13]. Based on the self-healing theories of inter-molecular diffusion and flow, the three factors coordination self-healing viewpoint was described to analyze the self-healing evolution behavior of bituminous materials with temperature [14, 15].

The second class of self-healing model could be classified as *physical–chemical-based numerical model*, which was also applied to explore the self-healing mechanisms by simulating the microscopic healing process, including the phase-field self-healing model and the molecular dynamics (MD) self-healing model. Kringos et al. [16] studied the microscopic multi-phase nature of bitumen based on atomic force microscope (AFM) investigations and developed a phase-field self-healing model. It could present the micro-scale process of crack initiation and disappearance by phase transition [17–19]. Moreover, MD model simulated the self-healing process of bitumen micro-cracks from the molecular scale point of view. It had achieved great progress in the bitumen crack self-healing investigation with good precision [20–22].

The third type could be classified as *mechanical-based self-healing model*, which was used to predict the self-healing performance by developing a damage or fracture self-healing constitutive model. This chapter presents the relevant research details and progress of the current three types of self-healing models. Moreover, the development and future modeling outlook of the self-healing process in bituminous materials would be discussed.

2 Physical–Chemical-Based Experimental Model

2.1 Inter-molecular Diffusion Self-healing Model

2.1.1 Inter-diffusion at Crack Faces

The self-healing mechanism based on inter-molecular diffusion come from the polymer chain dynamics theory of polymer system. De Gennes [3] put forward a reptation model to describe the self-diffusion of molecules moving in vermicular form in crosslinked polymer gels. In 1981, Wool and O'Connor [2] depicted the self-healing behavior of polymer from the perspective of inter-molecular diffusion. The self-healing process in the surfaces included five continuous phases: rearrangement, proximity, wetting, diffusion, and randomization [2, 23–26]. The model was shown in Fig. 1. In short, the model defined the self-healing process through the recovery of secondary bonds between molecules or microstructures, and the generation of force-transfer bridges on the interface through Rouse diffusion or reptation [3].

These phases would lead to close interface contact between the faces, thus restoring the inherent strength [2]. It was found that the rearrangement effect may

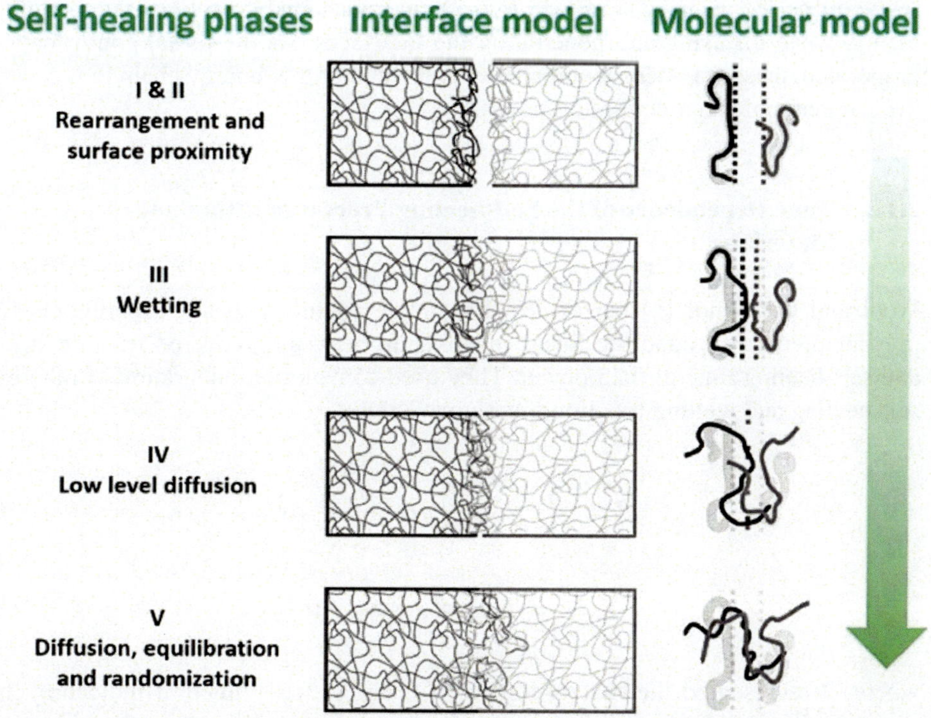

Fig. 1 Schematic representation of inter-molecular diffusion based self-healing phases. Modified from Qiu [26]

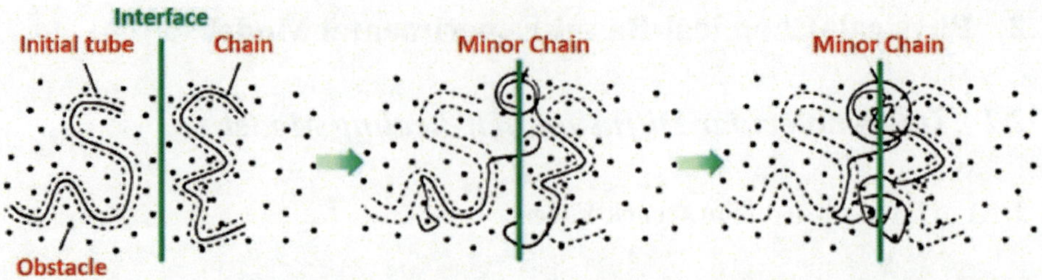

Fig. 2 Molecule separation from the initial conformation in the interfaces. Modified from Sun et al. [31]

lower the interface strength [27]. In the proximity phase, when the molecular fragment had adequate energy to cross the energy barrier [27] or driving force, part of the mechanical contact on the crack surface occurred through the molecular flow [27]. Wetting could remove the barrier related to the inhomogeneity, and the polymer chain could diffuse on the crack surface [27], enhancing the interface strength. The long-range interaction (electrostatic) of attraction between the non-bonded molecular segments was restored, thereby driving the adsorption. The inter-diffusion and randomization of chains on the surfaces were due to the entanglement and interpenetration between molecules [28]. Kim and Wool [29] determined the escaped segments as secondary chains (see Fig. 2), and presumed a new Gaussian conformation when diffusing and penetrating the interfaces. As the initial conformation evaporated, these fragments entangled with the end of low-energy chain to recovery the original molecular configuration [30].

2.1.2 Time Dependence of the Self-healing Process in Bituminous Materials

Wool and O'Connor [2] performed the double cantilever beam experiments for polybutadiene and found the linear relationship between recovered fracture stress and self-healing time of 0.25 power. They used convolution integration of inherent self-healing and wetting functions, as shown below:

$$R = \int_{\tau=\infty}^{\tau=t} R_h(t-\tau)\frac{\mathrm{d}\varphi(t, X)}{\mathrm{d}\tau}\mathrm{d}\tau \tag{1}$$

$$R_h(t) = R_0 + K \cdot t^{0.25} \cdot \phi(t) \tag{2}$$

where R represented the macroscopic self-healing effect function indicating the wetting and strength gain [4]. $R_h(t)$ indicated the intrinsic self-healing effect function including the instantaneous strength from cohesion (R_0) and the time-dependent strength from inter-molecular diffusion at faces (K was constant). $\varphi(t, X)$ was the wetting effect function, and τ was the time variable. $\phi(t)$ represented the effect of molecular rearrangements.

Based on the work conducted by Wool and O'Connor in polymer self-healing [2], investigators put forward related theories to apply to bituminous materials. Someone would point out that the long chains entanglements may not occur in bitumen. Additionally, Kim et al. [32] proved the significance of the bitumen molecules structure to self-healing. Bhasin et al. [4] built a fracture self-healing model, describing the time dependences of wetting process and intrinsic self-healing process. Sun et al. [6] determined the temperature-dependent self-healing process of bituminous mastic on the basis of inter-molecular diffusion theory.

The precursor of crack propagation at viscoelastic medium was the development of fracture zone at cracking tip [8]. Fracture process zone could be composed of multiple micro-cracks or nano-cracks. The cracks healed at once after the loading was removed. Figure 3 showed the focus area of crack/self-healing process [4]. The length of the crack/self-healing process zone, which played a role in the self-healing process through the inter-molecular force on the crack surfaces, was expressed as β. The tensile stress between crack faces was expressed in σ_b. The crack closing speed rate in the self-healing process zone was expressed as $\dot{a}_b$.

Wetting ability was related to the material property of bitumen. The surface free energy (or cohesive work), viscoelasticity, and fracture toughness of bitumen determines the wetting rate of two cracks. Bitumen with greater surface free energy had stronger inter-molecular attraction between two crack faces, so it would wet faster. The wetting step in the crack surfaces can be expressed by the wetting distribution function in convolution integral. Schapery put forward the relationship between cohesive work and material property relevant to wetting [8]. According to the relationship, the wetting distribution function or wetting rate (i.e., crack closing speed) was shown as:

$$\frac{d\varphi(t, X)}{dt} = \dot{a}_b = \beta\left[\frac{1}{D_1 k_m}\left\{\frac{\pi W_c}{41 - \nu^2\sigma_b^2\beta} - D_0\right\}\right]^{-1/m} \tag{3}$$

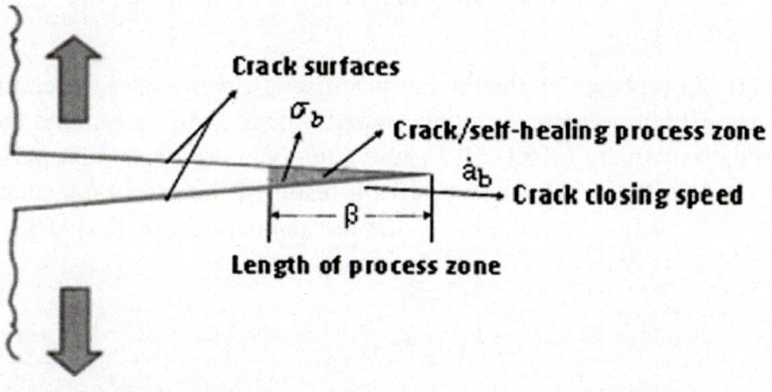

Fig. 3 Crack/self-healing process zone. Modified from Little and Bhasin [33]

where W_c was the cohesive work; ν denoted the Poisson's ratio; D_0, D_1, and m were acquired through fitting $D(t) = D_0 + D_1 \cdot t^m$; and k_m represented a material constant related to m. σ_b, β, and $\dot{a}_b$ had been explained in Fig. 3.

The intrinsic self-healing step mainly included the interfacial cohesion resulted strength gain driven by surface free energy and the inter-molecular diffusion between wetted surfaces. Bhasin et al. [4] and Bommavaram et al. [5] proposed the intrinsic self-healing function $R_h(t)$ of bitumen based on the modified Avrami equation, see Eq. (4).

$$R_h(t) = R_0 + p\left(1 - e^{-qt^r}\right) \tag{4}$$

where $R_h(t)$ was the time-dependent function, which represented the promotion of mechanical properties at the wetted crack surfaces. It was usually percentage of the same properties of the undamaged material and could be applied to denote any mechanical properties. R_0 denoted the instantaneous self-healing effect of wetted surfaces as result of interfacial cohesion. The right half in Eq. (4) denoted the dynamics of the molecular diffusion and randomization at the faces. q and r denoted the time-independent constants related to the molecular weight and activation energy, which could be determined via dynamic shear rheometer (DSR).

2.1.3 Temperature Dependence of the Self-healing Process in Bituminous Materials

Ishort, the higher the temperature, the shorter the self-healing recovery time. Concretely, temperature changed the wettability through the nucleation density and expansion rate of the wetting region, the diffusion rate and molecular mobility, and the chains penetration depth through the conformational effect [2]. Sun et al. put forward the macro recovery function of bituminous materials Eq. (5) [6], including the influences of self-healing time and temperature:

$$HI(t, T) = HI_0(T) + D(T) \cdot t^{0.25} \tag{5}$$

where $HI(t, T)$ represented the overall macroscopic self-healing strength recovery effect at the test temperature T and measured time t. $HI_0(T)$ denoted the instantaneous strength recovery effect. $D(T)$ was a temperature-dependent parameter and could be used as the self-healing rate as a result of the inter-molecular diffusion from one crack surface to the other at the test temperature T [14]. $D(T)$ could be represented by the Arrhenius law of diffusion as follows:

$$D(T) = D_0 \exp\left(-\frac{E_h}{RT}\right) \tag{6}$$

where D_0 denoted the equation constant, R denoted the universal gas constant (8.314 J/mol K), and T denoted the test temperature in degrees K. E_h was the activatio. energy required for time-dependent strength gain, representing the lowest energy needed for the molecular diffusion and randomization between the crack faces. Moreover, E_h can be also regarded as the self-healing activation energy, varying with the bitumen type and the damage levels. Through the substitution of Eq. (6) into Eq. (5), the overall macroscopic self-healing function was expressed as:

$$HI(t,T) = HI_0(T) + D_0 \exp\left(-\frac{E_h}{RT}\right) \cdot t^{0.25} \quad (7)$$

According to Eq. (7), the self-healing strength gain of bituminous materials mainly included two parts, i.e., the instantaneous part and time-dependent part, as depicted in Fig. 4. In the work done by Sun et al. [6, 22], the fatigue–healing–fatigue (FHF) experiments via DSR at various temperature and time were conducted to establish Eq. (7). Theoretically, if there was a considerable energy equivalent to or higher than the self-healing activation energy at the damaged surfaces, the time-varying self-healing recovery would be activated, that is, self-healing reaction. Otherwise, the reaction would stop. Therefore, the energy index, like E_h, could be developed as a stable indicator to assess the self-healing level of bitumen.

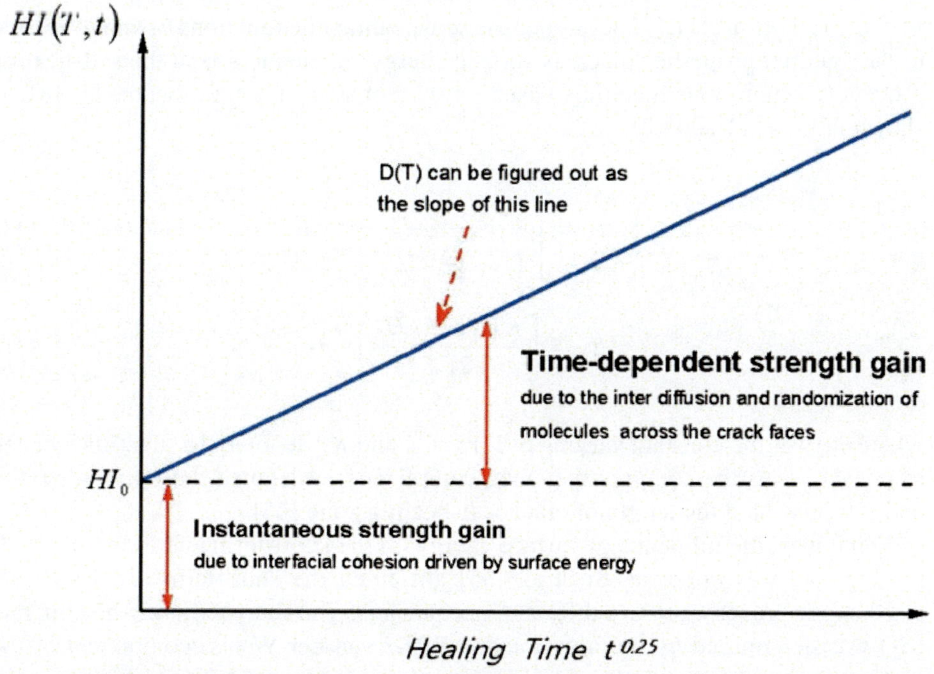

Fig. 4 Hypothesis of self-healing process depending on the time. Modified from Sun et al. [6]

2.2 Surface Energy Self-healing Model

According to the theory of fracture mechanics, the crack self-healing process of bituminous materials was manifested as the change of surface area, that is, the reduction of crack surface energy. Lytton et al. [7] and Little et al. [34] developed the basic laws of self-healing in bituminous materials Eqs. (8 and 9), which referred to the fundamental fracture mechanics rules in the crack process of viscoelastic media proposed by Schapery, as exhibited in Eq. (10) [8]:

$$2\Gamma_h = E_R D_h(t_\alpha) H_v \tag{8}$$

$$D_h(t) = D_{0h} + D_{1h} t^{m_h} \tag{9}$$

$$2\Gamma_f = E_R D_f(t_\alpha) J_v \tag{10}$$

where Γ_f and Γ_h represented the fracture surface energy density and self-healing surface energy density at cracking face, respectively. E_R was the reference modulus, $D_f(t_\alpha)$ and $D_h(t_\alpha)$ denoted the tensile creep compliance and compressive creep compliance at time t_α, J_v was the viscoelastic J integral, and H_v was the viscoelastic H integral. m_h was the slope of creep compliance with logarithmic time.

Further, Lytton et al. [7] developed the self-healing rate functions $\dot{h}_1$ and $\dot{h}_2$ related to the material properties (such as surface energy) to characterized the self-healing process in bituminous materials based on the Schapery fracture mechanics [8], as shown in Eqs. (11 and 12).

$$\dot{h}_2 = \left[\frac{2 r_m E_R^2 D_{1h} \Gamma_h}{(1 - \nu^2) C_m^{\frac{1}{m_h}} H_v} \right] \beta \tag{11}$$

$$\dot{h}_1 = \left[\frac{K_h E_R D_{1h} H_v}{2\Gamma_h} \right]^{\frac{1}{m_h}} \beta \tag{12}$$

where m_h was the constant varying 0–1. r_m, C_m and K_h denoted the functions related to m_h, D_{1h} was the compressive creep compliance, ν denoted the Poisson's ratio, and β represented the length of crack self-healing zone [35].

Obviously, the influence of surface energy on the short-term and long-term self-healing rate was opposite. Si et al. [35] attributed the short-term and long-term self-healing processes to two different mechanisms. The short-term self-healing rate ($\dot{h}_1$) was determined by the non-polar (Lifshitz van der Waals) component of the surface energy density, and the long-term self-healing rate ($\dot{h}_2$) was driven by the polar (Lewis acid–base) component. The author meanwhile put forward the total self-healing rate $\dot{h}$ to describe the actual self-healing rate in bituminous mixture, as exhibited in Eqs. (13–15).

$$\dot{h} = \dot{h}_2 + \frac{\dot{h}_1 - \dot{h}_2}{1 + \frac{\dot{h}_1 - \dot{h}_2}{h_\beta}(\Delta t)_h} \quad (13)$$

$$\dot{h}_2 = \left[\frac{2r_m E_R^2 D_{1h} \Gamma^{AB}}{(1 - \nu^2) C_m^{\frac{1}{m_h}} H_\nu} \right] \beta \quad (14)$$

$$\dot{h}_1 = \left[\frac{K_h E_R D_{1h} H_\nu}{2\Gamma^{LW}} \right]^{\frac{1}{m_h}} \beta \quad (15)$$

where $(\Delta t)_h$ denoted the rest duration between the applied loading, and h_β varied ranging from 0 to 1 and represented the maximum extent of self-healing that could be reached by bitumen.

The proposed self-healing behavior equation related the self-healing rate of bituminous materials to material properties, and the model parameters could be identified by the uniaxial cyclic tensile fatigue experiment and surface energy experiment. The short-term self-healing rate was mainly dependent on $\dot{h}_1$ associated with H_ν / Γ^{LW}. The long-term self-healing rate slowed down to near $\dot{h}_2$, depending on Γ^{AB}/H_ν, as presented in Fig. 5. Si et al. analyzed the $\dot{h}_1$ and $\dot{h}_2$ of 12 types of bituminous mixtures [35]. Results indicated that the short-term self-healing rate varied significantly, the $\dot{h}_1$ of tested bituminous mixtures varied between 4.2 and 12.2; and the change range of $\dot{h}_2$ was much lower, varying from 0.06 to 0.15.

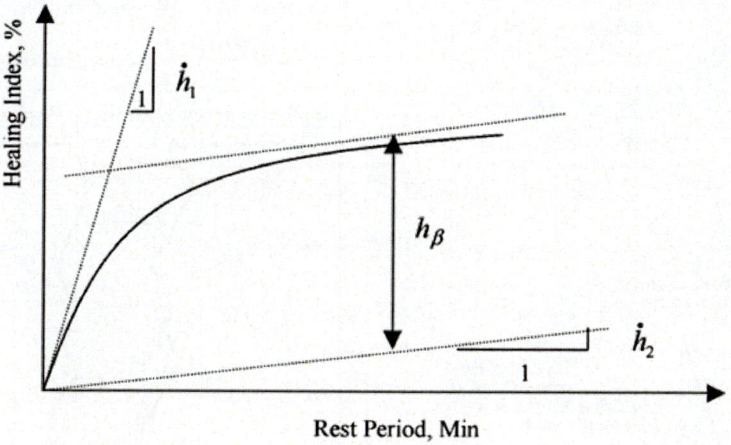

Fig. 5 Curve of self-healing index versus rest period. Modified from Sun et al. [31]

2.3 Capillary Flow Self-healing Model

Evidently, the self-healing process in bituminous materials was determined by the surface energy of materials and the inter-molecular diffusion and randomization abilities between the crack faces. Nevertheless, the inter-molecular diffusion self-healing model and surface energy self-healing theory were established based on the nano/micro-cracks that were invisible, and the intrinsic self-healing effects based on the inter-molecular diffusion or interaction at the crack surfaces were limited.

For the new macro-scale cracks inside bituminous materials (with structural damages), the inter-molecular diffusion would not happen spontaneously because of the large width between the crack surfaces. Nevertheless, if sufficient energy was used for the bituminous materials (such as through induction heating technology) and certain rest duration was allowed, the self-healing behavior would continue actively. Hence, the capillary flow self-healing theory was developed to analyze the meso-scale crack or macro-scale crack self-healing process through modifying the Lucas–Washburn equation [11, 12], as depicted in Fig. 6. The hypotheses of this model were put forward: at the appropriate temperature, the bitumen would be iquefied and flow into the cracks because of capillary force, and the inter-molecular diffusion would further promote the gradual recovery of interfacial strength.

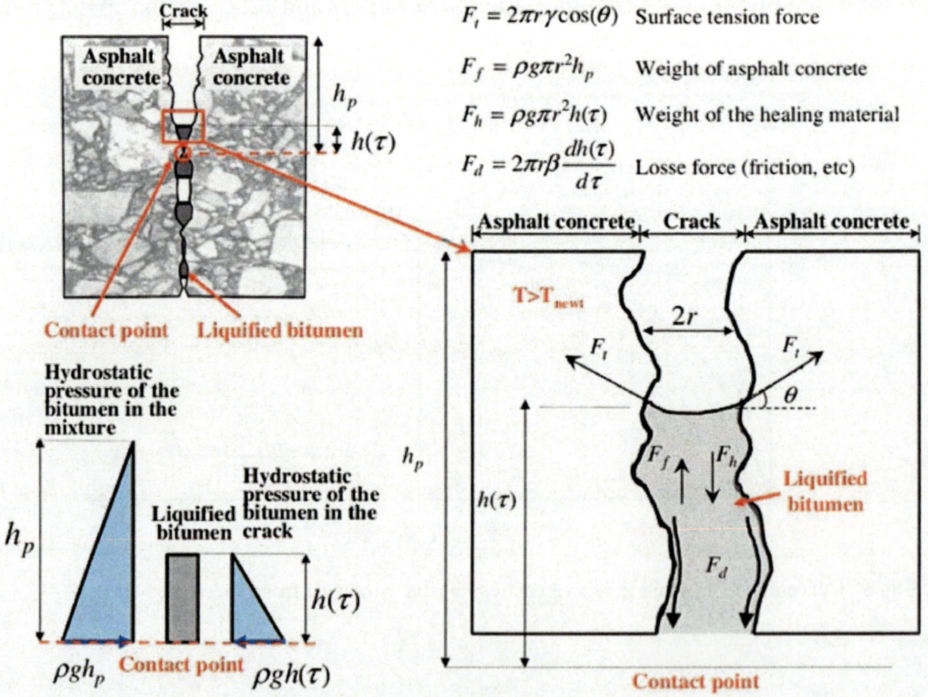

Fig. 6 Main forces occurring in the bituminous concrete crack self-healing process at higher temperature. Reproduced from García et al. [12]

The crack could be simplified as circular and vertical. Consequently, the main forces acting on the crack were produced as result of the surface tension (F_t), see Eq. (16).

$$F_t = 2\pi r\gamma\cos(\theta) \tag{16}$$

where r represented the capillary radius or half of crack width, γ denoted the surface tension of the iquefied bitumen, and θ was the contact angle in the wetting faces, which varied with the self-healing rate and the material types.

Besides, the force applied by gravity (F_g) could be regarded as the hydrostatic force at the zone, as exhibited in Eq. (17).

$$F_g = F_f - F_h = \rho g\pi r^2\left(h_p - h(\tau)\right) \tag{17}$$

where $\rho g\pi r^2 h_p$. (F_f) denoted the positive hydrostatic force in crack zone, $\rho g\pi r^2 h(\tau)$ (F_h) was the weight of the healing materials which depends on τ associated with the self-healing time and temperature. In addition, β was used to indicate the dissipated energy in the self-healing process, like the friction force of moving bitumen against the crack surfaces (F_d):

$$F_d - 2\pi r\beta\frac{\mathrm{d}h(\tau)}{\mathrm{d}\tau} \tag{18}$$

Therefore, the equilibrium equation with surface tension, gravity, and dissipative force was established as follows:

$$2\pi r\left(\gamma - \beta\frac{\mathrm{d}h(\tau)}{\mathrm{d}\tau}\right) = \rho g\pi r^2\left(h(\tau) - h_p\right) \tag{19}$$

$$h(\tau) = \left(h_p + \frac{2\gamma}{\rho g r}\right)\left(1 - \mathrm{e}^{-\frac{\rho g r\tau}{2\beta}}\right) \tag{20}$$

The occurrence speed of self-healing process was constant at all directions. After a period of self-healing, supposing that the number of possible contact points (expressed as n) in the self-healing process was certain, the effective area $A_E(\tau)$ of the cracked beam will be:

$$A_E(\tau) = \int_0^\tau 2\pi n h(\tau) dh(\tau) = C\cdot\mathrm{e}^{-D\tau}\left(-1 + \mathrm{e}^{\frac{D\tau}{2}}\right)^2 \tag{21}$$

$$C = \pi n\left(h_p + \frac{2\gamma}{\rho g r}\right)^2 \tag{22}$$

$$D = \frac{\rho g r}{\beta} \tag{23}$$

It should be noted that the necessary condition of capillary flow self-healing model was that the bitumen stayed in Newtonian fluid state. However, at lower temperature (usually lower than 40 °C), bitumen becomes a non-Newtonian fluid, and thus, the capillary flow model would no longer be applicable.

2.4 Three Factors Coordination Self-healing Viewpoint

Sun et al. [14, 15] concluded that the self-healing power of bitumen was mainly composed of flow, wetting diffusion, and elastic recovery, and these three factors promoted self-healing synergistically. Three factors are significantly affected by temperature within a wide temperature range [14, 22, 36]; thus, a recent study confirmed that the self-healing mechanism of bitumen evolved with temperature. Based on MD simulation, Gao et al. [36] pointed out that the self-healing behavior of bio-oil regenerated bitumen included viscous flow and elastic recovery, and it depended on viscous flow at high temperature and elastic recovery at low temperature. It was well-known that the self-healing of neat bitumen was determined by its flow and diffusion ability. Generally, the softer the bitumen was, the stronger the flow and diffusion ability was.

Furthermore, Sun et al. [14] put forward the phenomenon of self-healing transition, that is, the order of self-healing performance between different bitumens would change before and after the self-healing cross temperature. They further found that SBS modified bitumen had self-healing dual effect, that is, the SBS structure in SBS modified bitumen promoted self-healing at lower temperature (self-healing was dominated by elastic recovery at lower temperature), while at higher temperature, the flow of bitumen phase was hindered by SBS structure (flow diffusion was dominant with the increase of temperature). This dual effect become more significant with the increase of the SBS content, until the self-healing platform temperature zone appeared when the content was 6%, as shown in Fig. 7.

The self-healing transition phenomenon and self-healing dual effect revealed the complexity of self-healing mechanism of SBS modified bitumen evolution with temperature. The three factors coordination self-healing viewpoint combined the molecular diffusion self-healing theory and flow self-healing theory, which well explained the self-healing evolution behavior of SBS modified bitumen with temperature. However, more research was needed to prove its validity, such as in bituminous mixture, and theoretical equation was expected to be developed, thus accurately describing the self-healing evolution behavior of bituminous materials with temperature.

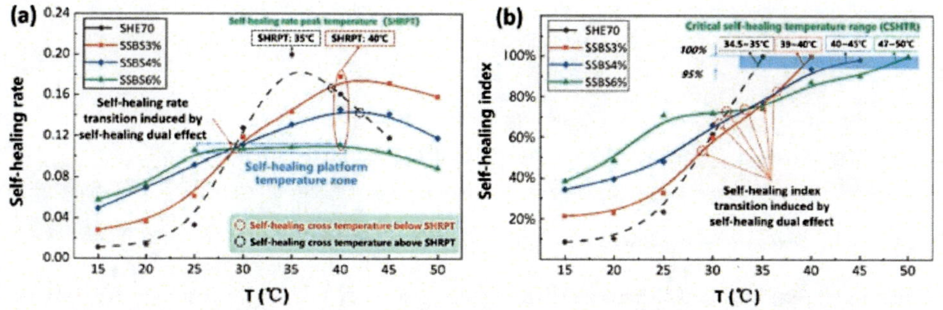

Fig. 7 Self-healing evolution behavior of bitumens with temperature: **a** Self-healing rate verse temperature curve; **b** Self-healing index verse temperature curve. Modified from Sun et al. [14]

3 Physical–Chemical-Based Numerical Model

3.1 Phase-Field Self-healing Model

The classic inter-molecular diffusion self-healing model was initially developed in the polymers [2, 4], while there were no long polymer molecular chains in bitumen. Hence, the proposed diffusion model was not fully applicable to the self-healing process of bituminous materials. As for the surface energy self-healing model, this theory regarded the bituminous materials as a continuous medium for research. But actually, the internal components and microstructure of bitumen were very complex. Thereby, there were some limitations in the analysis of self-healing mechanism of bitumen only from the macro-perspective. Hence, the phase-field self-healing model was developed to analyze the microscopic self-healing process together with the advanced AFM observation.

As shown in Fig. 8, Pauli [37] and Kringos et al. [16] established the connection between micro-phase separation and self-healing in bitumen based on AFM technology. Pauli used micro-finite element method (using CAPA 3D software platform) and AFM phase image to simulate the stiffness response of five various two-phase structures at monotonic load and cyclic load [37]. Moreover, it was observed that there was a remarkable difference in the internal phase of the bitumen during fatigue damage and healing. Therefore, the cracking and healing processes can be analyzed according to the bitumen's internal phase change. Results exhibited that there was a phase discontinuity between the domain phase and the surrounding matrix phase, thus resulting in the stress concentration, especially at the interface between various phases.

It has been extensively observed that two micro-phases would be separated in cooling and reach equilibrium state in heating [14, 37]. They diffused and coexist with each other. However, their stiffness and adhesion characteristics were different, meaning the inhomogeneity of bitumen microstructure. When subjected to load, the interface between the phases may appear stress concentration because of the

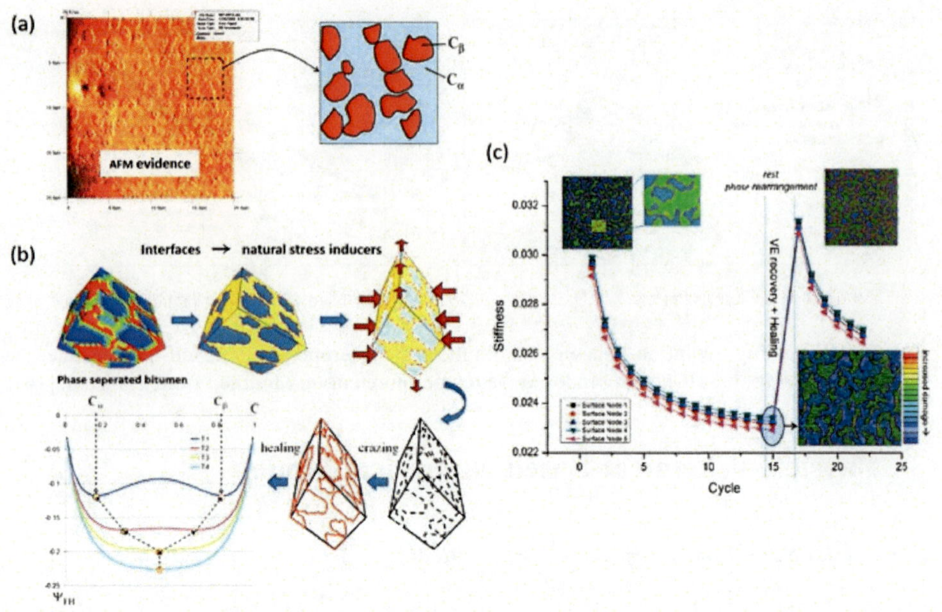

Fig. 8 Phase-field self-healing model in bitumen: **a** AFM phase separation in bitumen; **b** multi-phase healing model; **c** Internal phase change during damage healing. Reproduced from Kringos et al. [16] and Pauli [37]

different stiffness of each phase, eventually resulting in the formation of micro-cracks or crazes (i.e., cracking precursors) [38], see Fig. 9. Then, the micro-cracks propagated and coalesced into visible crack damage. Nevertheless, the development of micro-cracks could be stopped by eliminating the crazing in the microstructure.

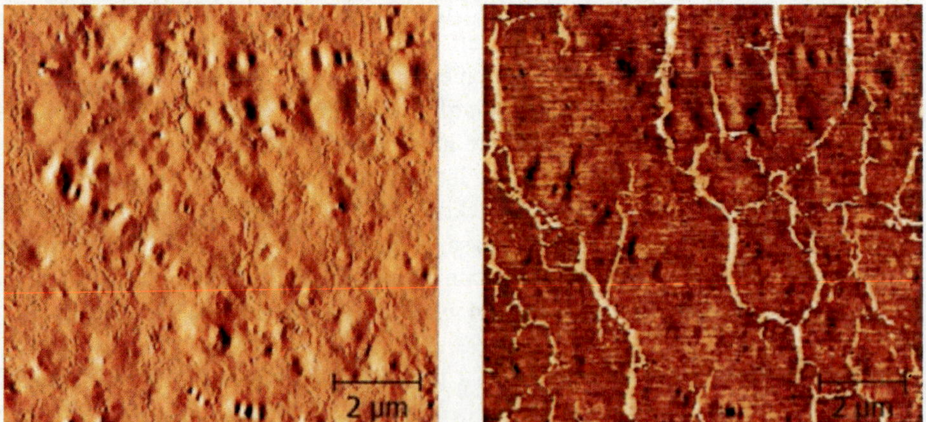

Fig. 9 Micro-cracks observed by AFM. Reproduced from Das et al. [38]

Through applying thermal energy or mechanical energy, the thermodynamic conditions of bitumen changed, resulting in phase rearrangement. The micro-phases will be arranged in a uniform mixing phase (i.e., single phase). By the phase transformation of microstructure, the memories of micro-crack or craze were eliminated, and the original microstructure performance of asphalt would be restored. Hence, the self-healing occurred after the above physical and chemical processes [16]. Furthermore, Nahar [39] investigated the micro-damage process of bitumen microstructure through various loading and found that micro-cracks initiated at the phase interfaces and propagated inward the domain phase at moderate loading condition; however, at high loading condition, micro-cracks coalesced and led to fragmented domain phases. Further, the cracks were gradually restored after moderate heating.

Kringos et al. [40] simulated the phase separation process of three SHRP bitumens based on the Cahn–Hilliard conservative phase-field model, see Fig. 10. Although the model did not analyze the bitumen microstructure, the precise simulation of the microscopic phase evolution of the different tested bitumens was achieved by adjusting the Cahn–Hilliard parameters. Subsequently, Hou et al. [18, 41–43] established a binary, ternary, and quaternary phase-field model from the perspective of thermodynamics and mechanics based on the phase-field theory using COMSOL Multiphysics software, and conducted a systematic study on the microscopic mechanical behavior, crack initiation, and healing process of bitumen. Table 1 exhibited the establishment process of the phase-field model [18, 41–43]. It could be summarized that micro-cracks were more likely to occur at the interfaces as a result of stress

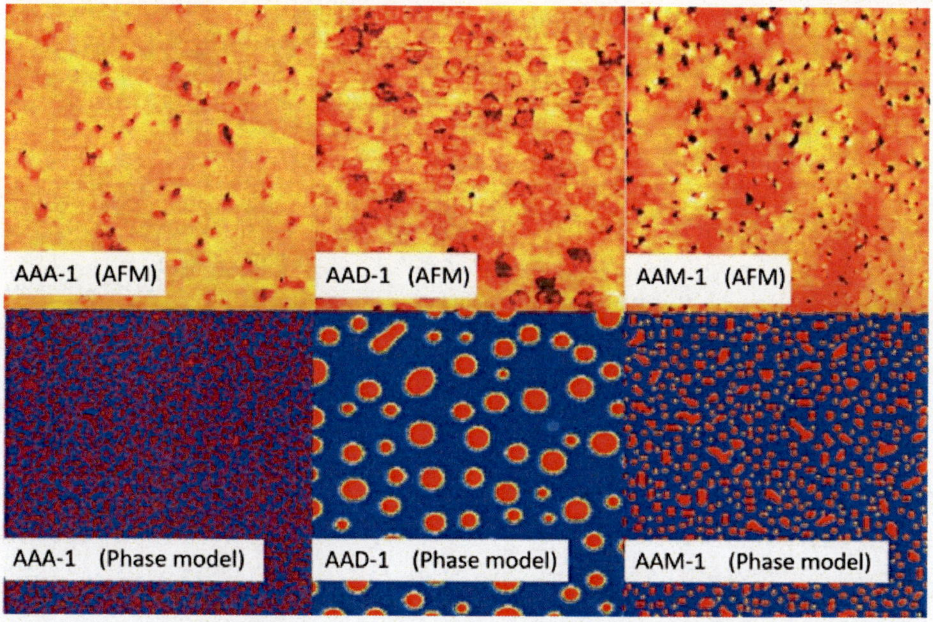

Fig. 10 Comparison of phase separation in AFM and phase-field models. Modified from Kringos et al. [40]

Table 1 Phase-field model establishment process from Hou et al. [18, 42, 43]

Types	Phase-field variable	Free energy function expression	Phase-field dynamic equation
Binary conservative phase-field model [18]	φ indicates the mass fraction of the different phases	$F = \int_{\Omega} \left(\begin{array}{l} \frac{1}{2}\kappa\lvert\nabla\phi\rvert^2 + \frac{1}{4\xi}\left(1-\phi^2\right)^2 \\ + \frac{1}{2}(\varepsilon-\varepsilon_0) : [E(\phi) : (\varepsilon-\varepsilon_0)] \end{array} \right)$	Cahn–Hilliard equation: $\frac{\partial\phi}{\partial t} = \nabla \cdot \left[M\nabla\frac{\delta F}{\delta\phi}\right]$
Binary non-conservative phase-field model [18]	$\phi = \pm 1$ represents the complete phase and the crack phase	$F = \int_{\Omega} \left(\begin{array}{l} \frac{1}{2}\kappa\lvert\nabla\phi\rvert^2 + \frac{1}{4\xi}\left(1-\phi^2\right)^2 \\ + \frac{1}{2}(\varepsilon-\varepsilon_0) : [E(\phi) : (\varepsilon-\varepsilon_0)] \end{array} \right)$	Allen–Cahn equation: $\frac{\partial\phi}{\partial t} = -M\frac{\delta F}{\delta\phi}$
Ternary conservative phase-field model [42]	ϕ_1, ϕ_2, ϕ_3 indicate the mass fraction of asphaltenes, naphthene aromatics, and saturates in bitumen	$F = \phi_1 \ln\phi_1 + \phi_2 \ln\phi_2 + \phi_3 \ln\phi_3$ $+ \kappa_{12}\phi_1\phi_2 + \kappa_{13}\phi_1\phi_3 + \kappa_{23}\phi_2\phi_3$	Cahn–Hilliard equation: $\frac{\partial\phi_i}{\partial t} + \mu\cdot\nabla\phi_i = \nabla\cdot\left[M\nabla\left(\frac{\partial F}{\partial\phi_i} - \nabla\cdot\frac{\partial F}{\partial(\nabla\phi_i)}\right)\right]$
Quaternary conservative phase-field model [43]	$\phi_1, \phi_2, \phi_3, \phi_4$ indicate asphaltenes, resins, oil, and wax mass fraction in bitumen	$F = \phi_1 \ln\phi_1 + \phi_2 \ln\phi_2 + \phi_3 \ln\phi_3 + \phi_4 \ln\phi_4$ $+ \kappa_{12}\phi_1\phi_2 + \kappa_{13}\phi_1\phi_3 + \kappa_{14}\phi_1\phi_4$ $+ \kappa_{23}\phi_2\phi_3 + \kappa_{24}\phi_2\phi_4 + \kappa_{34}\phi_3\phi_4$	Coupling of Cahn–Hilliard equation with Navier–Stokes equation $\rho\left(\frac{\partial u}{\partial t} + u\cdot\nabla\mu\right) = -\nabla p + \nabla\cdot\eta\left[\nabla\mu + (\nabla\mu)^{\mathrm{T}}\right] + \rho\left(\begin{array}{l} \mu_1\nabla\phi_1 + \mu_2\nabla\phi_2 \\ +\mu_3\nabla\phi_3 + \mu_4\nabla\phi_4 \end{array}\right)$

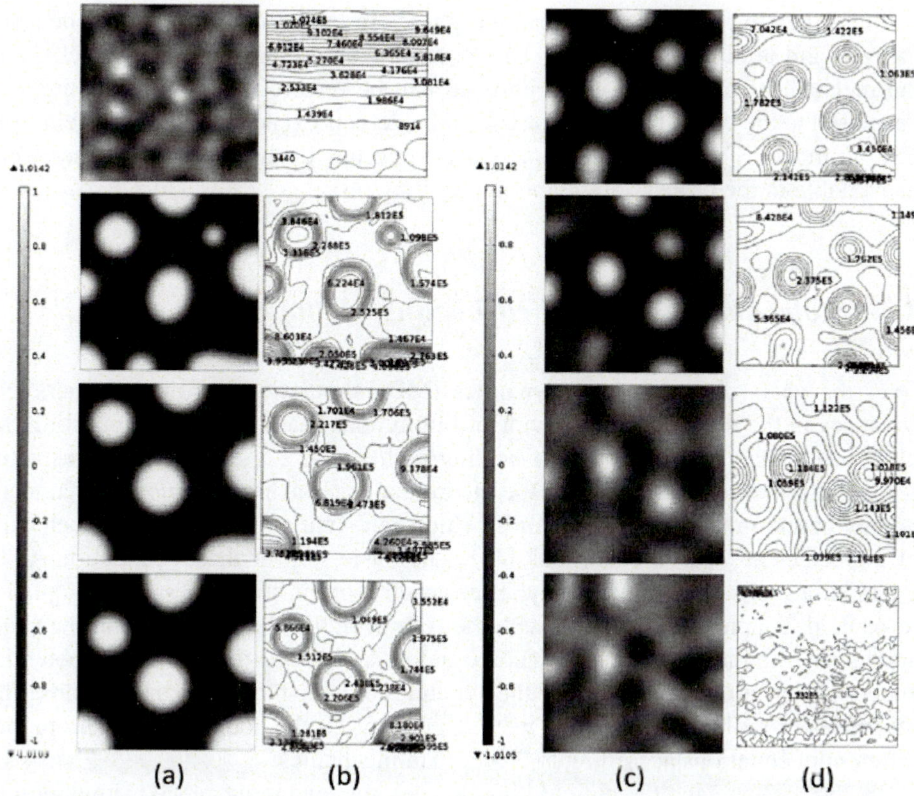

Fig. 11 **a**, **b** With the decrease of temperature, phase separation and the corresponding stress distribution; **c**, **d** With the increase of temperature, phase mixing and the corresponding stress distribution. Reproduced from Hou [41]

concentration and disappear during the rising process of temperature, thus leading to crack healing of bitumen.

Figure 11 showed the simulation results of the micro-phase change and stress distribution of bitumen based on the thermodynamic phase-field model [18]. It was seen that the phase separation not only affected the microstructure of bitumen, but also affected the distribution of internal stress of bitumen, mainly manifested in: (i) before and after phase separation, the internal stress distribution was relatively gentle, while in the process of phase separation, the internal stress fluctuated greatly [43]; (ii) during phase separation, there was stress concentration area at the junction of two phases. Under the action of temperature load, micro-cracks were first produced in stress concentration area and macro-fatigue cracks were formed continuously. This further confirms the observation results of AFM shown in Fig. 9. Meanwhile, the simulation results indicated that with the increase of temperature, the stress distribution in bitumen tended to be flat. The "bee structure" phase gradually melted into the continuous phase. The stress concentration phenomenon disappeared, and the micro-cracks gradually shrink and finally self-heal.

In the heating process, the separated phases in bitumen surface would be reorganized and tend to be isotropic. This process would erase the micro-cracks, and the mechanical properties of the bitumen can be restored when temperature was decreased to the initial level. At present, some defects existed in the phase-field self-healing model. For example, the physical properties of the separated phases were unavailable in the process of the thermal cycling test.

3.2 *Molecular Dynamics Self-healing Methods*

Over the years, the molecular dynamics (MD) simulation has made remarkable advances in the self-healing behavior of bituminous materials, thus supporting the multi-scale analysis results of self-healing ability [21, 22, 31, 44]. MD was actually a collective term, referring to the theoretical approach and computational technology based on statistical mechanics, which can simulate the interaction behavior of molecules at various conditions [45]. The basic step of these simulations is to establish a molecular model firstly. Then, each atom in molecules was assigned a force field, which was consistent with the type of molecular structure and the characteristics of adjacent atoms [44]. Force field was applied to identify the potential energy of a group of molecules and the dynamics of inter-molecular interactions. The properties of molecular ensemble were dependent on the coupled nonlinear partial differential equations according to Newtonian mechanics.

In 2007, Zhang and Greenfield [46–48] took the lead in using MD simulation to estimate the performance of model bitumen. They combined the various molecular models (including asphaltenes, saturated hydrocarbons, naphthenic aromatics, and polar aromatics) into a unit cell to prepare various representative binders. Then in 2011, Bhasin et al. [44] first employed MD simulation for bitumen self-healing analysis. The diffusion coefficient was put forward to quantify the molecular motion state in the self-healing process. Results further confirmed the relationships between the molecular chain size, the branching degree and the inter-molecular self-diffusion coefficient, thus affecting the microscopic self-healing rate of bitumens. Further, Shen et al. [49] established the MD self-healing models with various micro-crack widths to study the micro-mechanical self-healing mechanisms of bitumen via the open-source software LAMMPS (large-scale atomic/molecular massively parallel simulator). It was showed that the diffusion mechanism of bitumen molecules triggered the self-healing, and the higher the temperature, the higher the molecular diffusion rate (i.e., self-healing rate). Additionally, Xu and Wang [21] built a two-layer bitumen self-healing model with 10 Å (10^{-10} m) micro-crack layer. The inter-molecular diffusion potential and activation energy of aged bitumen and unaged bitumen at various temperatures. Results indicated that the aged bitumen had smaller diffusion capacity and greater self-healing activation energy barrier than original binder.

Regarding the self-healing process of bitumen-additive system, the MD simulation provided insights into the molecular interaction mechanism of self-healing. Sun et al. [20] constructed the three-component molecular models of base bitumen,

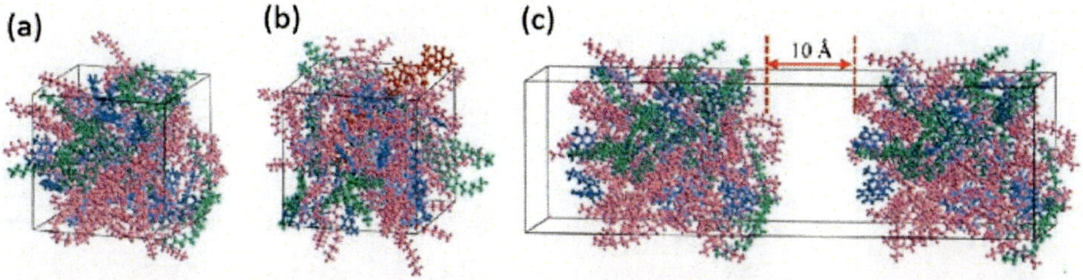

Fig. 12 **a**, **b** Built molecular model cell of base bitumen and SBS modified bitumen; **c** Bitumen crack-healing model for MD simulation. Modified from Sun et al. [20]

SBS modified bitumen model, and crack-healing molecular model, see Fig. 12. They analyzed three indices obtained by MD simulation results, i.e. the diffusion coefficient, activation energy and pre-exponential factor, and verified their reliability in characterizing the self-healing properties of bitumens. He et al. [50] also investigated the molecular diffusion of SBS modified bitumen, virgin bitumen, and aged bitumen, and concluded that the aging of bitumen molecules decreased the diffusion speed, and SBS additive could promote the diffusion speed of bitumen model. Moreover, Hung et al. [51] determined the molecular mechanism of self-healing and hindering effect of wax on self-healing in bitumens using the MD analysis and AFM observation. They innovatively used the density functional theory (DFT) via the $DMol^3$ module embedded in Accelrys Materials Studio software to analyze the role of wax in bitumen self-healing interface. Wax molecules migrated to the crack surface and form "bee" structure, and the surface roughness produced by "bee" structure prevents the conformal contact between crack surfaces, thus reducing the self-healing ability of bitumen.

Notably, Sun and Wang [52] proposed a novel simulated failure–healing–failure (FHF) test based on the MD simulation. They studied the crack development and strength recovery in the FHF process of bituminous binder under various conditions, see Fig. 13. Results reflected that the wetting and diffusion phases could be observed: (1) the fast recovery stage with short time, i.e., the wetting phase; and (2) the slow recovery stage, i.e., the surface diffusion phase. Furthermore, the self-healing ratios based on the cohesive strength were influenced by self-healing time, temperature, and damage level. Another novel finding was the influence of temperature on the self-healing properties of bituminous materials [22, 53]. Sun et al. [22] analyzed the inherent temperature sensitivity of self-healing property in bitumens through MD simulation. Results reflected that the inter-molecular diffusion speed and range are more prominent, especially in the phase transition temperature range. Yu et al. [53] found that the temperature self-healing process of bitumen nano-cracks had three phases, i.e., initial turbulent phase, distance self-healing phase, and strength self-healing phase.

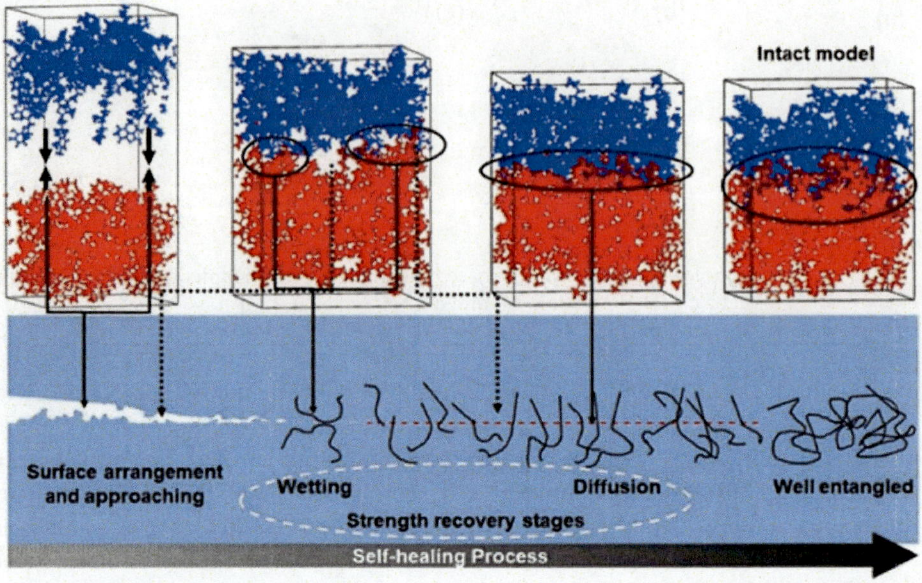

Fig. 13 Cracked bitumen molecular structures in various self-healing phases (blue and red molecules denoted two cracked fragments). Reproduced from Sun and Wang [52]

4 Mechanical-Based Self-healing Model

Over the years, the constitutive damage model coupled with the self-healing had attracted extensive attentions and become a new research field. Actually, a lot of researchers had been dedicated to the development of thermodynamic-based models for simulating viscoelastic, viscoplastic damage, and self-healing behaviors of different materials, such as metals and polymers [54–56]. In 2005, Barbero et al. [57] first put forward a continuum damage-healing model. They built an elasto-plastic damage self-healing constitutive equation to analyze the self-healing behavior in polymer composites. Most of the theoretical researches on the continuum models of damage-healing were based on the principle of continuum thermodynamics or the framework of continuum damage mechanics.

Numerical method had strong advantage in simulating the self-healing constitutive model, which can directly exhibit the continuous mechanical process of bituminous material consisting of damage and self-healing. Al-Rub et al. [58] put forward a micro-damage self-healing model, which combined with the temperature-dependent nonlinear viscoelastic, viscoplastic, and viscodamage constitutive models could anticipate the highly nonlinear mechanical response of bituminous mixture under the action of repeated loads. Furthermore, the corresponding codes based on the ABAQUS program called "Pavement Analysis Using Nonlinear Damage Approach (PANDA)" were developed [58]. In this damage-healing model, the effective stress concept was applied, see Fig. 14.

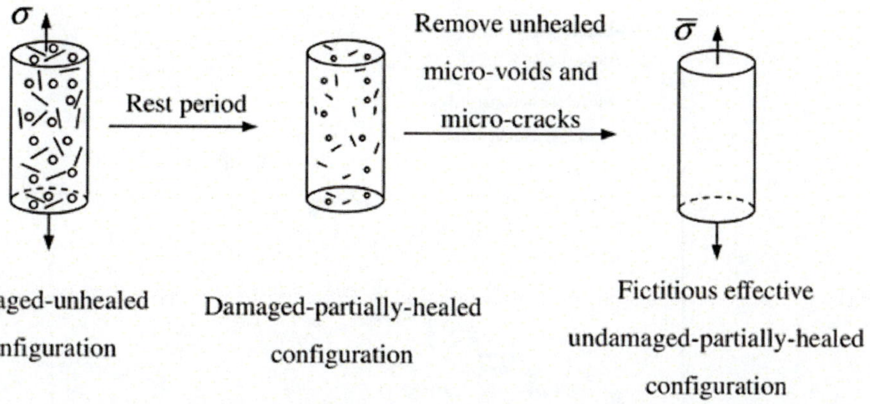

Fig. 14 Application of the concept of effective stress in continuum damage mechanics for damage-healed material. Reproduced from Al-Rub et al. [58]

In addition, inspired by the proposed self-healing model based on inter-molecular diffusion [2, 4], a phenomenological self-healing function for the evolution of internal state variables $\dot{h}$ of micro-damage self-healing was derived, which was related to rest time, temperature, damage degree, and self-healing history. Thereby, all these effects can be incorporated into the following healing evolution, Eq. (24).

$$\dot{h} = (1 - \overline{\varphi})^{b_1}(1 - h)^{b_2}\Gamma_0^h \exp\left[-\delta_3\left(1 - \frac{T}{T_0}\right)\right] \tag{24}$$

where $\dot{h} = dh/dt$ represented the self-healing rate; $\overline{\varphi}$ indicated the effective damage density; Γ_0^h denoted the self-healing viscosity parameter that determined how fast the material heals at the reference temperature of T_0; δ_3 was the self-healing temperature coupling parameter; T was the self-healing temperature; b_1, b_2 represented the material constants.

In practice, the self-healing related material parameters were calibrated and derived through modeling the repeated creep-recovery experiment, as exhibited in Fig. 15. More details of this mechanical based damage-healing modeling approach could be obtained in Al-Rub et al. [58]. Further, Darabi et al. [59, 60] developed a novel framework based on the continuum damage mechanics to model the micro-damage self-healing process in bituminous materials. In this framework, the concepts of effective configuration and stress were extended to self-healing configuration, and the analytical relations of three famous transformation hypotheses in continuum damage mechanics, i.e. strain, elastic strain energy, and power equivalent hypothesis, are given.

Furthermore, Luo et al. [61, 62] developed an energy-based mechanistic (EBM) method to carry out mechanical analysis on the self-healing behavior and model self-healing through the evolution of damage density. It is a purely mechanical approach based on a continuum concept called distributed continuum fracture (DCF) and

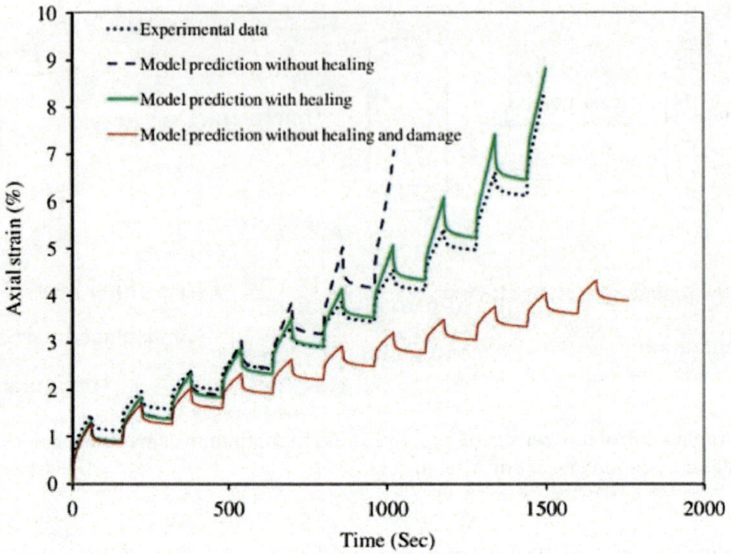

Fig. 15 Repeated creep-recovery test with 60 s load and 100 s rest each cycle: test creep strain and model-predicted strains with and without self-healing. Reproduced from Al-Rub et al. [58]

two basic mechanics principles (force equilibrium and energy balance). It was a pure mechanical method based on continuum concepts, which was called distributed continuum fracture (DCF) and two fundamental mechanical principles (force and energy balances). The EBM approach was applied to the creep and step-loading recovery (CSR) experiment to model healing in bituminous mixtures, thus quantifying the self-healing degree occurring in the rest time. More details can be found in Luo et al. [61]. The updated study from Luo et al. [63] further analyzed the self-healing kinetics in bituminous mixture based on the EBM method. They investigated the relationship between the different coordinated processes (i.e., free energy change and bond recovery) during self-healing and revealed that the self-healing rate was directly proportional to these two process rates, see Fig. 16.

In recent years, the damage or fracture mechanics was recommended to predict the self-healing influence on the fatigue damage evolution of the bituminous materials. Together with the computed tomography (CT) scanning technology, the real and complete 3D geometry model of the bituminous materials could be reconstructed in the numerical software, including the void distribution, aggregate contact points, and cracking expansion path. Therefore, more accurate mechanical responses coupling self-healing can be obtained by applying the damage-healing constitutive model to bitumen mortar phase (the aggregate phase can be considered as a linear elastomer).

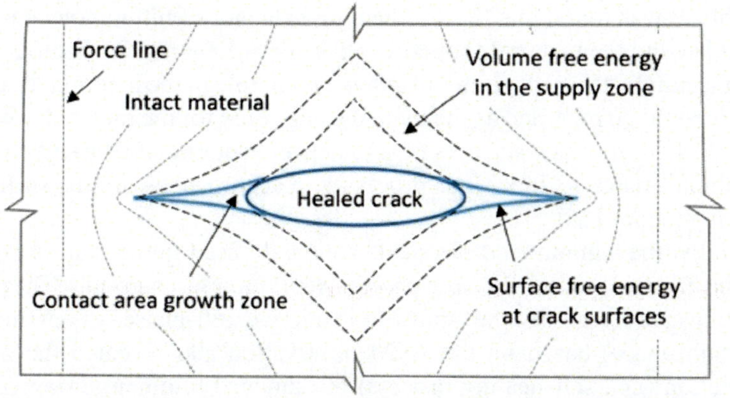

Fig. 16 Illustration of crack faces closure and bonds restoration. Reproduced from Luo et al. [63]

5 Summary and Outlook

Overall, it can be concluded from this chapter that the intrinsic self-healing behavior in bituminous materials was a very complex spontaneous process. This process was not only dependent on the material properties but also affected by the self-healing temperature, self-healing time, self-healing history, damage extent, and load condition. Extensive investigations built various models to assess the self-healing behavior of bituminous materials on the basis of various theories, and these works all pointed out clear goals and existing challenges.

According to different experimental methods and material types, the various physical–chemical-based experimental models were proposed to understand the self-healing behavior of bituminous materials. The inter-molecular diffusion self-healing model could characterize the self-healing of bituminous materials at the molecular level. This model could describe the self-healing developing stages of bituminous material, while it was hard to calibrate the model parameters associated with the crack zone and material properties. In contrast, the surface energy self-healing model could connect the short-term self-healing rate and long-term self-healing rate with various surface energy components (i.e., Lifshitz–Van der Waals and Lewis acid/base surface energy). Nevertheless, this theory was unable to reveal the development process of self-healing over time. Moreover, the capillary flow self-healing model could characterize the macro-scale crack self-healing in bituminous concrete, whereas this model required the Newtonian state of bitumen. It cannot assess the self-healing behavior of brittle bitumen at the lower temperatures (usually below 40 °C). As for three factors coordination self-healing theory, it could well interpret the self-healing evolution behavior with temperature, especially for SBS modified bitumen. Besides, it should be noted that the establishment of the experimental model usually depended on the test method, and the method change would lead to the change of the model form. For example, Xie et al. [64] developed a novel self-healing framework through the linear amplitude sweep-based self-healing test.

Notably, it was found that the self-healing behavior of bituminous materials was closely related to the material properties (like surface energy, or rheological properties). For example, the surface energy played great role in the mentioned self-healing models. It could be regarded as the main driving force for the molecular wetting and capillary flow in the crack surfaces [4, 11, 12], and determined the self-healing rates of bituminous materials in various processes in accordance with the surface energy self-healing theory [35].

Generally, the summarized physical–chemical-based numerical models characterized the microscopic self-healing process of bituminous materials. The phase-field self-healing model regarded the bitumen as a micro-multi-phase system (matrix phase and domain phase) based on the AFM observation and revealed the microstructure evolution (i.e., self-healing process) of damaged bitumen during cooling and heating according to the phase transformation theory. The MD simulation revealed the self-healing behavior from the molecular scale, while it was difficult to relate to macroscopic self-healing performance.

Moreover, the mechanical-based self-healing models were proposed to characterize the overall mechanical performance change with/without rest period under various conditions through regarding the bituminous materials as a continuum. The mechanical models can not only be connected with the experimental models, but also can predict the performance development of bituminous pavement throughout the service duration, so as to further optimize the design of pavement materials and structures, which was the main development direction of the self-healing model in the future. Nevertheless, there was yet a lot of work to be done for the improvement of the constructed damage or fracture constitutive model considering self-healing characteristics, the optimization of the calibration approaches, and the solution algorithm. As bituminous mixture is a complicated viscoelastic composite, there were lots of factors affecting its self-healing property, and some influencing mechanisms may be still not clear at present [58]. Therefore, there were imperative demands to develop the optimized numerical algorithm to quickly solve the damage self-healing constitutive model. Moreover, the coupling of environmental factors (like moisture damage and aging) was necessary because the final purpose was to predict the responses of bituminous pavement in its service period through modeling.

Acknowledgements The work described in this chapter was supported by Natural Science Foundation of China (Nos. 52108390, 52178434, 51878500), State Key Laboratory of High-Performance Civil Engineering Materials (No. 2021CEM010), Scientific Research Project of Beijing Educational Committee, China Postdoctoral Science Foundation (No. 2021M690271), Beijing Postdoctoral Research Foundation and Basic Research Project of Beijing University of Technology.

References

1. Kim YR, Little DN, Lytton RL (2003) Fatigue and healing characterization of bitumen mixtures. J Mater Civ Eng 15:75–83
2. Wool RP, O'Connor KM (1981) A theory crack healing in polymers. J App Phys 52:5953–5963
3. De Gennes PG (1971) Reptation of a polymer chain in the presence of fixed obstacles. J chem phys 55(2):572–579
4. Bhasin A, Little DN, Bommavaram R, Vasconcelos K (2008) A framework to quantify the effect of healing in bituminous materials using material properties. Road Mater Pav Des 9:219–242
5. Bommavaram R, Bhasin A, Little DN (2009) Determining intrinsic healing properties of bitumen binders: role of dynamic shear rheometer. Trans Res Rec 2126:47–54
6. Sun D, Lin T, Zhu X, Cao L (2015) Calculation and evaluation of activation energy as a self-healing indication of bitumen mastic. Constr Build Mater 95:431–436
7. Lytton RL, Uzan J, Fernando EG, Roque R, Hiltunen D, Stoffels SM (1993) Development and validation of performance prediction models and specifications for bitumen binders and paving mixes. Strategic Highway Research Program, Washington, DC
8. Schapery RA (1989) On the mechanics of crack closing and bonding in linear viscoelastic media. Inter J Frac 39:163–189
9. Little DN, Lytton RL, Williams D, Chen CW (2001) Microdamage healing in bitumen and bitumen concrete, vol I, Microdamage and microdamage healing, Project summary report
10. Little DN, Bhasin A (2008) Exploring mechanism of healing in bitumen mixtures and quantifying its impact. In: Self healing mater. Springer, Dordrecht, pp 205–218
11. García A (2012) Self-healing of open cracks in bitumen mastic. Fuel 93:264–272
12. García A, Bueno M, Norambuena-Contreras J, Partl MN (2013) Induction healing of dense asphalt concrete. Constr Buil Mater 49:1–7
13. Hamraoui A, Nylander T (2002) Analytical approach for the Lucas-Washburn equation. J Coll Interf Sci 250:415–421
14. Sun G, Hu M, Sun D, Deng Y, Ma J, Lu T (2020) Temperature induced self-healing capability transition phenomenon of bitumens. Fuel 263:116698
15. Sun G, Ma J, Sun D, Yu F (2021) Influence of weather accelerated ageing on healing temperature sensitivity of asphalts. J Clean Prod 281:124929
16. Kringos N, Scarpas A, Pauli T, Robertson R (2009) A thermodynamic approach to healing in bitumen. Taylor and Francis Group, London
17. Pauli AT (2014) Chemomechanics of damage accumulation and damage-recovery healing in bituminous bitumen binders. Phd. Thesis, TU Delft University of Technology
18. Hou Y, Wang L, Pauli T, Sun W (2014) Investigation of the bitumen self-healing mechanism using a phase-field model. J Mater Civ Eng 27:04014118
19. Nahar SN (2016) Phase-separation characteristics of bitumen and their relation to damage-healing. Delft University of Technology, TU Delft
20. Sun D, Lin T, Zhu X, Tian Y, Liu F (2016) Indices for self-healing performance assessments based on molecular dynamics simulation of bitumen binders. Com Mater Sci 114:86–93
21. Xu G, Wang H (2017) Molecular dynamics study of oxidative aging effect on bitumen binder properties. Fuel 188:1–10
22. Sun D, Sun G, Zhu X, Ye F, Xu J (2018) Intrinsic temperature sensitive self-healing character of bitumen binders based on molecular dynamics simulations. Fuel 211:609–620
23. Kline DB, Wool RP (1988) Polymer welding relations investigated by a lap shear joint method. Poly Eng Sci 28:52–57
24. Garcia SJ (2014) Effect of polymer architecture on the intrinsic self-healing character of polymers. Euro Poly J 53:118–125
25. Gaskin J (2013) On bitumen microstructure and the effects of crack healing. Phd. Thesis, University of Nottingham
26. Qiu J (2012) Self healing of bitumen mixtures: towards a better understanding of the mechanism. Phd. Thesis, TU Delft University of Technology

27. Wool RP (1983) Crack healing in polymers. Illinois University at Urbana Department of Metallurgy and Mining Engineering
28. Wool RP (1995) Polymer interfaces: structure and strength. Carl HanserVerlag, New York
29. Kim YH, Wool RP (1983) A theory of healing at a polymer-polymer interface. Macromolecules 16:1115–1120
30. Wool RP (1980) Crack healing in semicrystalline polymers, block copolymers and filled elastomers. Springer, Boston, MA, Adhe Adsor Poly, pp 341–362
31. Sun D, Sun G, Zhu X, Guarin A, Li B, Dai Z, Ling J (2018) A comprehensive review on self-healing of bituminous materials: mechanism, model, characterization and enhancement. Adv Coll Interf Sci 256:65–93
32. Kim YR, Little DN, Benson FC (1990) Chemical and mechanical evaluation on healing mechanism of bitumen concrete. J Assoc Asph Pav Tech 59
33. Little DN, Bhasin A (2007) Exploring mechanism of Healing in asphalt mixtures and quantifying its impact. In: Self-healing mater. Springer, Dordrecht, pp 205–218
34. Little DN, Prapnnachari S, Letton A, Kim YR (1993) Investigation of the microstructural mechanism of relaxation and fracture healing in bitumen. Texas Transportation Inst College Station
35. Si ZS, Little DN, Lytton RL (2002) Evaluation of fatigue healing effect of asphalt concrete by pseudostiffness. Trans Res Rec 1789:73–79
36. Gao X, Liu Z (2019) Self-healing mechanism of bio-oil recycled bitumen. China J Highway 32(4):235–242
37. Pauli AT (2014) Chemomechanics of damage accumulation and damage-recovery healing in bituminous bitumen binders. Delft University of Technology, TU Delft
38. Das PK, Jelagin D, Birgisson B, Kringos N (2012) Micro-mechanical investigation of low temperature fatigue cracking behaviour of bitumen. In: 7th RILEM inter confer cracking pave. Springer, Dordrecht, pp 1281–1290
39. Nahar SN (2016) Phase-separation characteristics of bitumen and their relation to damage-healing. PhD. Thesis, TU Delft University of Technology
40. Kringos N, Schmets A, Scarpas A, Pauli T (2011) Towards an understanding of the self-healing capacity of bitumenic mixtures. Heron 56:49–79
41. Hou Y (2014) Computational analysis of bitumen binder based on phase field method. Virginia Polytechnic Institute and State University
42. Hou Y, Wang L, Wang D, Guo M, Liu P, Yu J (2017) Characterization of bitumen micro-mechanical behaviours using AFM, phase dynamics theory and MD simulation. Mater 10:208
43. Hou Y, Sun W, Das P, Song X, Wang L, Ge Z, Huang Y (2016) Coupled Navier-Stokes phase-field model to evaluate the microscopic phase separation in bitumen binder under thermal loading. J Mater Civ Eng 28:04016100
44. Bhasin A, Bommavaram R, Greenfield ML, Little DN (2011) Use of molecular dynamics to investigate self-healing mechanisms in bitumen binders. J Mater Civ Eng 23:485–492
45. Frenkel D, Smit B (2001) Understanding molecular simulation: from algorithms to applications. Academic Press
46. Zhang L, Greenfield ML (2007) Analyzing properties of model bitumens using molecular simulation. Ener Fuels 21:1712–1716
47. Zhang L, Greenfield ML (2007) Relaxation time, diffusion, and viscosity analysis of model bitumen systems using molecular simulation. J Chem Phys 127:194502
48. Zhang L, Greenfield ML (2008) Effects of polymer modification on properties and microstructure of model bitumen systems. Ener Fuels 22:3363–3375
49. Shen S, Lu X, Liu L, Zhang C (2016) Investigation of the influence of crack width on healing properties of bitumen binders at multi-scale levels. Constr Buil Mater 126:197–205
50. He L, Li G, Lv S, Gao J, Kowalski KJ, Valentin J, Alexiadis A (2020) Self-healing behavior of asphalt system based on molecular dynamics simulation. Constr Build Mater 254:119225
51. Hung AM, Mousavi M, Fini EH (2020) Implication of wax on hindering self-healing processes in bitumen. App Surf Sci 523:146449

52. Sun W, Wang H (2020) Self-healing of asphalt binder with cohesive failure: Insights from molecular dynamics simulation. Constr Build Mater 262:120538
53. Yu T, Zhang H, Wang Y (2020) Multi-gradient analysis of temperature self-healing of asphalt nano-cracks based on molecular simulation. Constr Build Mater 250:118859
54. Kachanov L (2013) Introduction to continuum damage mechanics. Springer Science & Business Media
55. Schapery RA (1999) Nonlinear viscoelastic and viscoplastic constitutive equations with growing damage. Inter J Frac 97:33–66
56. Ghorbel E (2008) A viscoplastic constitutive model for polymeric materials. Inter J Plas 24:2032–2058
57. Barbero EJ, Greco F, Lonetti P (2005) Continuum damage-healing mechanics with application to self-healing composites. Inter J Dam Mech 14:51–81
58. Al-Rub RKA, Darabi MK, Little DN, Masad EA (2010) A micro-damage healing model that improves prediction of fatigue life in bitumen mixes. Inter J Eng Sci 48:966–990
59. Darabi MK, Al-Rub RKA, Masad EA, Little DN (2012) Thermodynamic-based model for coupling temperature-dependent viscoelastic, viscoplastic, and viscodamage constitutive behaviour of bitumen mixtures. Inter J Nume Analy Meth Geomech 36:817–854
60. Darabi MK, Al-Rub RKA, Little DN (2012) A continuum damage mechanics framework for modeling micro-damage healing. Inter J Sol Struc 49:492–513
61. Luo X, Luo R, Lytton RL (2015) Mechanistic modeling of healing in bitumen mixtures using internal stress. Inter J Sol Struc 60:35–47
62. Luo X, Luo R, Lytton RL (2014) Energy-based mechanistic approach for damage characterization of pre-flawed visco-elasto-plastic materials. Mec Mater 70:18–32
63. Luo X, Birgisson B, Lytton RL (2020) Kinetics of healing of asphalt mixtures. J Clea Prod 252:119790
64. Xie W, Castorena C, Wang C, Kim YR (2017) A framework to characterize the healing potential of asphalt binder using the linear amplitude sweep test. Constr Build Mater 154:771–779

Self-adaptive Construction Materials: Future Directions

Antonios Kanellopoulos, Magdalini Theodoridou, Michael Harbottle, Sergio Lourenco, and Jose Norambuena-Contreras

1 Introduction

Self-healing construction materials have caused great interest in the last decade due to their several applications for civil engineering. A self-healing material is defined as an artificial or synthetically created material that has the built-in ability to repair damage to itself without human intervention. In this context, recent advancements in self-healing construction materials have opened new opportunities and pathways for the construction sector to promote a more resilient civil infrastructure.

Self-healing mechanisms have been developed, and proof-of-concept ideas have been tested from laboratory experimentation to large-scale application. At the same timeframe, remarkable advancements were made in the field of materials science.

A. Kanellopoulos (✉)
Division of Civil Engineering & Built Environment, School of Physics, Engineering & Computer Science, University of Hertfordshire, Hatfield, Hertfordshire, UK
e-mail: a.kanellopoulos@herts.ac.uk

M. Theodoridou
Hub for Biotechnology in the Built Environment, Newcastle University, Newcastle upon Tyne, England
e-mail: Magdalini.Theodoridou@newcastle.ac.uk

M. Harbottle
Cardiff School of Engineering, Cardiff University, Cardiff, Wales
e-mail: HarbottleM@cardiff.ac.uk

S. Lourenco
Department of Civil Engineering, The University of Hong Kong, Hong Kong, China
e-mail: lourenco@hku.hk

J. Norambuena-Contreras
LabMAT, Department of Civil and Environmental Engineering, University of Bío-Bío, Concepción, Chile
e-mail: jnorambuena@ubiobio.cl

© Springer Nature Switzerland AG 2022
A. Kanellopoulos and J. Norambuena-Contreras (eds.), *Self-Healing Construction Materials*, Engineering Materials and Processes,
https://doi.org/10.1007/978-3-030-86880-2_8

Polymerisation techniques were improved, new coating materials were developed, and the introduction of the graphene family compounds came to revolutionise the way materials are designed and used. The traditionally conservative construction sector comes gradually to the realisation that we cannot rely anymore on "traditional" materials that have remained almost unaltered for about a century. The construction industry needs more advanced materials with adaptive and multifunctional purposes that allow them to change their properties as response to the environmental and load-bearing stimuli they are exposed. Very often these days we hear the term 'smart cities' or 'cities of the future' and mainly the reference there is on artificial intelligence, electronic features and/or smart vehicles. Materials are downplayed in this discussion and that is a critical mistake. The cities of the future cannot be delivered with the conventional materials of the past.

This chapter aims to give some ideas for future directions focused on self-adaptive construction materials. Of course, the list below is not exhaustive, but it rather comes to complement existing ideas about the future outlook of smart construction materials than can be found in the literature.

2 Cement-Based Self-healing Materials

At the beginning of the twenty-first century, concrete and other cement-based materials were considered composites with specific properties and functionalities. Their wide availability and low cost led to be treated as consumable commodities that can be produced in vast volumes and replaced as often as needed. Twenty years later, we have a totally different situation upon us. Ensuring sustainability and protecting our natural resources are becoming more emergent than ever. At the same time, research and development in the field have progressed significantly. New cement-based composites have emerged, ranging from ductile ultra-high-performance fibre-reinforced composites to self-healing concrete. The latter has grown considerably within the last ten years, with significant developments from groups worldwide. An overview of these developments are discussed in Chap. "Self-healing Cement-Based Materials: Mechanisms and Assessment" while there are currently two comprehensive reviews in the published literature [1, 2].

The global interest on self-healing cement-based materials is also evident by the large number of research projects that received funding from leading research councils such as the Engineering and Physical Sciences Research Council (EPSRC) in the UK, the National Academy of Sciences in the USA, the Chinese Academy of Sciences and the European Commission. There were also regional development funds in different parts of the world that supported such research (e.g., funding councils in Belgium, Netherlands, Chile, etc.). In the UK, since 2013, EPSRC has funded, with almost £10 M in total, the materials for life and the resilient materials for life consortia [3] which have provided significant contribution in the state-of-the-art in the field [4, 5]. At wider continental level, the importance of self-healing and adaptive materials was recognised by the European Commission who provided support in two wider reaching and implementation actions: (i) Self-healing as prevention repair of concrete structures (SARCOS) and (ii) Self-healing multifunctional advanced repair technologies in cementitious systems (SMARTINCS) [6, 7]. However, besides the progress and wider interest, there are still several points of interest that need to be further explored.

2.1 Standardisation and Repeatability

Standardisation can be a significant barrier towards the wider implementation of self-healing systems in cement-based materials that concern standardisation of both materials used and assessment techniques. On the materials side, currently different research groups utilise different compounds and different approaches. There are some points of convergence (e.g., when it comes to specific minerals used), but there are also significant differences in terms of dosages, encapsulation materials and delivery methods. Similarly, a lot of work is needed in establishing some common experimental techniques that can efficiently assess healing levels. Current methods have been very useful in providing justification and support in proof-of-concept studies. However, as discussed in Chap. "Self-healing Cement-Based Materials: Mechanisms and Assessment", the methods used are not designed for assessing healing. In fact, some of them are even sensitive in the existence of cracks in the matrix (e.g., sorptivity; gas permeability). Therefore, there is a great challenge and opportunity for the scientific community to work on the development of new modified and widely acceptable assessment techniques. These should include both laboratory and field/large-scale assessment.

The aspect of repeatability of results is another issue that in some ways connects to standardisation as well. Although the number of studies on self-healing mechanisms in cement-based composites is significant, for many mechanisms, there is no proof of systematic repeatability of findings. To be clear, this does not necessarily mean that the systems do not work but it rather highlights further the need for widespread round-robin-tests. For example, when it comes to mineral additions the data presented in the literature shows significant widespread of findings that range from 'promoting considerable healing' to 'no significant effect on healing' [1]. Similar issues can be found when it comes to microcapsules, vascular networks and bacteria for example. Recently, through SARCOS action the initiation of some interlaboratory research was done but more is needed and for a variety of mechanisms [8].

2.2 *Healing Under Varying Conditions/Large-Scale Applications*

So far, studies on self-healing cement-based composites have been focusing on proving the functionality of specific proposed systems under controlled and largely steady-state conditions within the confinements of laboratories. Nonetheless, the implementation of such healing mechanisms in real structures requires more research to ensure that self-healing actions take place and maintained under common infrastructure operational conditions. One of the 'advertised' advantages of self-healing materials is that they can reduce or even eliminate disruption caused by traditional maintenance regimes. However, at present, the sector does not have enough or convincing evidence of that. Can self-healing mechanisms in a structural member of a highway bridge manifest themselves under continuous traffic, for example? If the bridge needs to shut down, what is the timeframe for this? Post-healing, can the healed members sustain the combined actions, mechanical and environmental, acting on the bridge? Can self-healing mechanisms work under pressure or temperature differentials? Can self-healing mechanisms function under multiple exposure conditions? From practical aspect, these are very important and critical questions that require detailed and convincing answers.

Another area of significant importance is the study of self-healing mechanisms on a large scale. It is very promising that recently few such investigations are reported in different parts of the world [9–12]. Nonetheless, these case studies need to expand and cover a variety of self-healing mechanisms, structures, exposure conditions and timespans. It is difficult to extract safe conclusions about long-term efficiency if a large-scale application lasts a limited period or covers only a narrow set of parameters. Assessing self-healing for, ideally, several years will allow better understanding of the long-term performance of self-healing concrete as well as the shelf-life of embedded healants.

3 Self-healing Bituminous Materials by Encapsulated Agents

Different technologies to promote the self-healing in bituminous materials (including asphalt mixtures, mastics and bituminous binders) have made significant advances in recent years. As discussed in Chap. "Advances in Self-healing Bituminous Materials: From Concept to Large-Scale Application", two approaches, from a concept in laboratory to on-site large-scale applications, have mainly been used to promote crack-healing in bituminous materials: (i) an extrinsic approach to reduce the viscosity of bitumen by increasing its temperature through externally triggered heating using magnetic field technologies, such as electromagnetic induction heating [13] and microwave radiation [14]; (ii) an autonomic crack-healing approach by means of releasing bitumen miscible encapsulated rejuvenating agents [15].

Research discussed in Chap. "Advances in Self-healing Bituminous Materials: From Concept to Large-Scale Application" has proved that induction and microwave heating technologies can recover the original mechanical properties of the samples and heal artificial cracks of different sizes in bituminous materials. Nevertheless, these heating approaches also present some disadvantages that often reduce the material's integrity, and do not help to solve the problem of the ageing of bitumen [13]. Therefore, in recent years, the concept of extrinsic self-healing by the action of encapsulated rejuvenating agents has been considered as a revolutionary future technology for the autonomic crack-healing of bituminous materials, with the additional potential to restore the original properties in the aged bitumen.

As Gonzalez-Torre and Norambuena-Contreras [16] discussed in their review article that includes more than 100 scientific papers on self-healing asphalt, researchers are carrying out intense work to promote the autonomic healing of bituminous materials by encapsulated rejuvenating agents. Several authors from countries such as China, United Kingdom, Netherlands, USA and Chile have successfully synthesised and proved the asphalt self-healing with different encapsulated rejuvenating agents derived from natural and man-made resources. Overall, researchers have focused their attention on the development of smaller particles to encapsulate rejuvenators, like core–shell microcapsules, polynuclear (multi-cavity) capsules and hollow or compartmented fibres. Among these developments in encapsulated rejuvenators, polymer-based polynuclear capsules are the most promising technique for asphalt self-healing with proved results. Finally, numerous tests including standard and innovative approaches have been reported to quantify the self healing capability of bituminous materials with encapsulated rejuvenators. In these tests, self-healing capability of bituminous materials has been measured under several laboratory conditions through the concept of the 'healing level' [16].

3.1 More Sustainable Encapsulated Materials for Asphalt Healing

Despite the urgent need to develop more sustainable construction materials, and the extensive diversity of renewable resource-based polymers for these purposes, the use of biomaterials in the civil engineering industry remains limited. Based on a circular economy vision for more sustainable and resilient asphalt pavements, current research has been mainly focused on the valorisation of waste as a promising resource to produce eco-friendly encapsulated rejuvenator for asphalt self-healing.

Recent studies have focused their efforts on the design and production of more sustainable microcapsules for asphalt self-healing, where renewable resource-based polymers obtained from biomass waste have been used as encapsulation materials for virgin and recycled rejuvenating agents. Studies to date have included, but are not limited to: (1) use of bio-oil obtained from liquefied agricultural biomass waste as a bio-based encapsulated rejuvenating agent for self-healing of bituminous materials

[17], (2) production of polynuclear biocapsules containing low-cost rejuvenators for asphalt self-healing [18], (3) evaluation of the self-healing capability of asphalt materials using polynuclear biopolymer capsules containing waste oils [19] and more recently (4) use of pyrolytic oil from waste tyres as an encapsulated rejuvenator to promote the extrinsic self-healing in bituminous materials [20].

Overall, these studies are focused on the design of economically viable solutions to self-heal asphalt by using more sustainable capsules, although future research must be developed to address the gaps on capsules for asphalt self-healing.

3.2 Future Research on Asphalt Healing by Encapsulated Agents

Based on the literature review of Gonzalez-Torre and Norambuena-Contreras [16], future research on self-healing technology by capsules should be focused on the following suggestions: (1) the design of new experimental tests that simulate the self-healing of bituminous materials by capsules in a more realistic manner; (2) the design and manufacturing of encapsulation up-scale devices to increase the capsule volume production to be included in large-scale applications; (3) the understanding of how capsules are activated inside the aged bituminous material, and which is the mechanism through which the rejuvenator heals the aged binder; (4) more research on the experimental and numerical multiscale modelling (as discussed in Chaps. "Multiscale Measurements of the Self-healing Capability on Bituminous Materials" and "Modelling of Self-healing Process in Bituminous Materials: Experimental and Numerical Models") that allows the definition of structure–function relationships for understanding the self-healing procedures using encapsulated agents; (5) the design of smart microcapsules that could be externally monitored by non-destructive techniques; and finally, (6) the development of more research from a multidisciplinary scientific, technical and economic approach, to increase the knowledge of the self-healing technology by capsules that promises to increase the lifespan of roads.

4 Biological Self-healing for Geological Construction Materials and Masonry Structures

Geological materials are used in significant quantities for construction, in infrastructure and comprise most heritage building materials worldwide. However, they are subject to damage, and deterioration through weathering, everyday stresses and other human activities [21–34] which could compromise the structural integrity and safety of buildings, as well as reduce their aesthetic appeal.

To prevent this from happening, or to reduce the rate at which it occurs, new structures may be overdesigned such that they take longer to reach a critical state, or existing structures may be subject to regular maintenance programs. Maintenance is disruptive and costly, but without consideration of material durability and deterioration (which may be difficult to observe or detect), there are significant safety and cultural implications for the long-term performance of such structures. Data availability is lacking, but an indication could be given through the example of a single city, Glasgow, where the masonry maintenance costs were estimated at £30 M per year [35]. The *GEOHEAL* project [36–38] explored for the first time the potential for providing such materials with the ability to self-heal via naturally occurring biological mechanisms—to automatically respond to damage or deterioration, instigating a process that restores or enhances the desirable structural or aesthetic properties.

4.1 Biological Self-healing for Existing Masonry and Historic Buildings

Providing masonry and stonework with an in-built immune system ready to work before damage becomes critical will significantly benefit their longevity and reduce their maintenance needs. In so doing, this approach promotes the adaptation and reuse of existing properties and heritage buildings as an alternative solution to the increasing development of new constructions, offering radical solutions that align with the international targets for sustainable development and reduced emissions. Also, it is worth noting that by incorporating a self-healing system into this type of sustainable building materials, we enhance their performance in a way that could play a vital role in reintroducing their wider use in modern construction.

Stone and geological materials are by nature bioreceptive and suitable for biomineralisation thanks to their mineralogical composition and porous microstructure [39–44]. Also, the healing products of the developed self-healing systems are well-matched to the original substrate [45]. For example, calcium-rich stones, such as limestones or sandstones, could be considered a better type of stone for biological self-healing with calcite (the most common biomineralisation product), due to the match between the healing product and the host material and the availability of calcium for the formation of new material. However, it is possible, and possibly even advantageous, to provide such precursor chemicals in order to boost the efficiency of the self-healing system and prevent decomposition of the natural fabric of the host material. Such a method has already been demonstrated to create a self-healing sandstone-type material from sand as a method of soil improvement [46]. Biological methods of healing have already been explored in some depth for use in cement-based materials [1, 47–49]. However, the principal difference is that the harsh environment in cementitious materials requires either extremophilic microorganisms adapted to high pH or protection of the bacteria from the alkaline environment. The bacteria used often originate from soils and are likely to find surviving and forming minerals in stone less of a challenge than in cementitious environments, as long as their basic needs are met (e.g., supply of water, oxygen, nutrients, precursor chemicals for mineralisation).

The *GEOHEAL* project [36] focused on the use of *Sporosarcina pasteurii* and *Sporosarcina ureae,* which are both non-hazardous, naturally occurring soil bacteria. The results proved that the developed biological systems could work and remain efficient after damage, even providing cyclic healing properties to the host materials. Current research at the *Hub for Biotechnology in the Built Environment* [50], apart from exploring the full potential of novel biotechnological processes and applications using naturally occurring bacteria for sustainable building and smart heritage conservation considers engineered bacteria systems for better tailoring the system to the needs of specific applications, for instance, when exposure to particularly harsh environmental conditions is expected or the aesthetics of the new or existing structure require modification of the healing product.

4.2 Bacteria Types, Activity and Application

Biological self-healing could be responsive at any depth as long as the bacteria have access to water/oxygen, which eventually takes place when a crack occurs, and the nutrients supply; that adds a considerable benefit to this mechanism, as most known mechanisms for protecting stone are usually restricted to surface coatings. Also, conventional surface protection measures often limit the 'breathability' of the material, leading to spalling and accelerated weathering of the material through its newly exposed surface. The biological treatments studied during the *GEOHEAL* project-induced sufficient cementation in the near surface region, which is the area more prone to weathering, to an extent that could be considered protective, yet compatible with materials' natural properties. The outcomes of the study also suggest that the system using *S. ureae* could survive harsh weathering conditions and reactivate, enabling cyclic healing, further highlighting the potential of the biological systems to benefit the long-term performance of geological construction materials and masonry structures, while reducing their maintenance needs.

4.3 Pilot Tests and Future Use

In order to deliver research impact through testing on heritage masonry materials, representative stone types from a heritage site in Wales, which exhibits masonry deterioration, were selected and subjected to experimentation which proved the efficiency of the developed biological system to change the microstructure of the materials exposed to weathering. Yet, the applied treatment did not seem to alter the treated zone of the stone specimens to an extent that could be proven incompatible with the substrate or compromise it aesthetically, for example, no colour alterations were found that could be detectable by the human eye. There is more work planned for testing and monitoring the long-term performance of biological self-healing in a historic but also contemporary context, in urban and rural environments, including the HBBE's experimental house—The OME [51].

Realistically speaking, and until these technologies become widely accepted and therefore more affordable, biological self-healing mechanisms are ideal for materials and structural elements which are either in need of repair after having experienced damage or are in critical or difficult-to-access locations and an early on treatment could prevent damage. In both cases, maintenance costs would be significantly reduced, and their extended service life could contribute to meeting the global sustainability goals. In new construction, the possibility of incorporating self-healing materials and structural elements would enable bolder and more sustainable designs, and that is also of great importance.

5 Application of Self-healing Approaches to Granular Materials

Self-healing approaches present an opportunity to switch or recover important properties in granular materials, from strength to wettability. Hydrophobized granular materials, for instance, can be used in ground infrastructure to delay water infiltration [52]. For slopes, to reduce pore water pressure built-up in natural or infrastructure slopes and other sloping grounds; for clay deposits, to reduce the risk of volumetric changes and desiccation cracks; for landfills and porous pavements; as a layered hydrophilic/hydrophobic system, to drain or store water; for water harvesting, to collect rainfall in arid regions; in food defences, to form a temporary barrier to flooding water; and for foundations, to minimize concrete degradation by aggressive pore water [53]. However, the lifespan of hydrophobicity in soils is variable, with previous studies reporting a range from a few months to decades.

The durability of the hydrophobic polymers used in construction materials is influenced by their intrinsic chemical properties and environmental factors such as water, sunlight, temperature, biological activities and particle abrasion [54]. The degradation is mostly related to oxidation and hydrolysis, depending on the polymer type. For instance, hydrolysis contributes to the deterioration of polyesters, aramids and polyamides during the application, while for polyolefins, it is mostly driven by oxidation. The application environment of hydrophobized granular materials, at the ground-atmosphere interface or buried, can accelerate the hydrolysis or oxidation and consequently affect their durability. Consequently, hydrophobicity being a surface property is unlikely to last the lifespan of an engineering structure. The level of hydrophobicity induced in these granular materials will eventually decline and their surfaces revert to the original that is hydrophilic state.

Micro-encapsulation approaches initially developed for self-healing applications emerge as a potential solution to enhance, switch (from hydrophilic) or prolong the longevity of hydrophobized granular materials. These approaches have been successfully implemented with cementitious construction materials. For concrete, the healing principle is based on the rupture of microcapsules by fissures which releases healants, leading to precipitation and proliferation of crystalline products filling the fissures. This results in healing of the cracks and recovery of significant properties such as permeability and strength. For granular materials, there is a need to test physicochemical methods to encapsulate and release hydrophobic substances in a porous environment. Therefore, any future research will need to consider aspects specific to granular materials (porous nature, environmental impact) which may differ from cementitious materials. This includes a variety of factors, from the type of cargo and shell, to soil pH and temperature including the microcapsules ecotoxicity, durability and strength.

References

1. De BN, Gruyaert E, Al-tabbaa A et al (2018) A review of self-healing concrete for damage management of structures. Adv Mater Interfaces 1800074:1–28. https://doi.org/10.1002/admi.201800074
2. Zhang W, Zheng Q, Ashour A, Han B (2020) Self-healing cement concrete composites for resilient infrastructures: A review. Compos B Eng 189:107892
3. RM4L resilent materials for life—RM4L. https://rm4l.com/. Accessed 12 May 2021
4. Teall OR, Pilegis M, Sweeney J et al (2017) Development of high shrinkage polyethylene terephthalate (PET) shape memory polymer tendons for concrete crack closure. Smart Mater Struct 26:045006. https://doi.org/10.1088/1361-665X/aa5d66
5. Al-Tabbaa A, Litina C, Giannaros P et al (2019) First UK field application and performance of microcapsule-based self-healing concrete. Constr Build Mater 208. https://doi.org/10.1016/j.conbuildmat.2019.02.178
6. SARCOS self-healing as prevention repair of concrete structures. https://www.sarcos.eng.cam.ac.uk/. Accessed 12 May 2021
7. SMARTINCS smartincs. https://smartincs.ugent.be/index.php/about-us. Accessed 12 May 2021
8. Litina C, Bumanis G, Anglani G et al (2021) Evaluation of methodologies for assessing self-healing performance of concrete with mineral expansive agents: an interlaboratory study. Materials (Basel) 14:2024. https://doi.org/10.3390/ma14082024
9. Davies R, Teall O, Pilegis M et al (2018) Large scale application of self-healing concrete: design, construction, and testing. Front Mater 5:1–12. https://doi.org/10.3389/fmats.2018.00051
10. Van MT, Gruyaert E, Caspeele R, De BN (2020) First large scale application with self-healing concrete in Belgium: analysis of the laboratory control tests. Materials (Basel) 13:997. https://doi.org/10.3390/ma13040997
11. Qian C, Zheng T, Zhang X, Su Y (2021) Application of microbial self-healing concrete: case study. Constr Build Mater 290:123226. https://doi.org/10.1016/j.conbuildmat.2021.123226
12. Wang X, Huang Y, Huang Y et al (2019) Laboratory and field study on the performance of microcapsule-based self-healing concrete in tunnel engineering. Constr Build Mater 220:90–101. https://doi.org/10.1016/j.conbuildmat.2019.06.017
13. Norambuena-Contreras J, Garcia A (2016) Self-healing of asphalt mixture by microwave and induction heating. Mater Des 106:404–414. https://doi.org/10.1016/j.matdes.2016.05.095
14. Norambuena-Contreras J, Gonzalez-Torre I (2017) Influence of the microwave heating time on the self-healing properties of asphalt mixtures. Appl Sci 7:1076. https://doi.org/10.3390/app7101076

15. Garcia A, Jelfs J, Austin CJ (2015) Internal asphalt mixture rejuvenation using capsules. Constr Build Mater 101:309–316. https://doi.org/10.1016/j.conbuildmat.2015.10.062
16. Gonzalez-Torre I, Norambuena-Contreras J (2020) Recent advances on self-healing of bituminous materials by the action of encapsulated rejuvenators. Constr Build Mater 258:119568
17. Norambuena-Contreras J, Arteaga-Perez LE, Guadarrama-Lezama AY et al (2020) Microencapsulated bio-based rejuvenators for the self-healing of bituminous materials. Materials (Basel) 13. https://doi.org/10.3390/ma13061446
18. Concha JL, Arteaga-Pérez L, Gonzalez-Torre I, Norambuena-Contreras J (2021) Biocapsules containing low-cost rejuvenators for asphalt self-healing. RILEM Tech Lett 6:1–7. https://doi.org/10.21809/rilemtechlett.2021.129
19. Yamaç ÖE, Yilmaz M, Yalçın E et al (2021) Self-healing of asphalt mastic using capsules containing waste oils. Constr Build Mater 270:121417. https://doi.org/10.1016/j.conbuildmat.2020.121417
20. Norambuena-Contreras J, Arteaga-Pérez LE, Concha JL, Gonzalez-Torre I (2021) Pyrolytic oil from waste tyres as a promising encapsulated rejuvenator for the extrinsic self-healing of bituminous materials. Road Mater Pavement Des. https://doi.org/10.1080/14680629.2021.1907216
21. Price CA, Amoroso GG, Fassina V (1984) Stone decay and conservation: atmospheric pollution, cleaning, consolidation and protection. Stud Conserv 29:158. https://doi.org/10.2307/1506020
22. Camuffo D Microclimate for cultural heritage. Science Direct. Springer
23. Modestou S, Theodoridou M, Fournari R, Ioannou I (2016) Physico-mechanical properties and durability performance of natural building and decorative carbonate stones from Cyprus. Geol Soc Spec Publ 416:145–162. https://doi.org/10.1144/SP416.3
24. Gibeaux S, Vázquez P, De Kock T et al (2018) Weathering assessment under X-ray tomography of building stones exposed to acid atmospheres at current pollution rate. Constr Build Mater 168:187–198. https://doi.org/10.1016/j.conbuildmat.2018.02.120
25. Mol L, Gomez-Heras M (2018) Bullet impacts and built heritage damage 1640–1939. Herit Sci 6:35. https://doi.org/10.1186/s40494-018-0200-7
26. Theodoridou M, Török Á (2019) In situ investigation of stone heritage sites for conservation purposes: a case study of the Székesfehérvár Ruin Garden in Hungary. Prog Earth Planet Sci 6:1–14. https://doi.org/10.1186/s40645-019-0268-z
27. Jokilehto J (2015) A history of architectural conservation, 2nd edn
28. Maravelaki-Kalaitzaki P, Biscontin G (1999) Origin, characteristics and morphology of weathering crusts on Istria stone in Venice. Atmos Environ 33:1699–1709. https://doi.org/10.1016/S1352-2310(98)00263-5
29. McAlister JJ, Smith BJ, Török A (2008) Transition metals and water-soluble ions in deposits on a building and their potential catalysis of stone decay. Atmos Environ 42:7657–7668. https://doi.org/10.1016/j.atmosenv.2008.05.067
30. Brimblecombe P, Grossi CM (2009) Millennium-long damage to building materials in London. Sci Total Environ 407:1354–1361. https://doi.org/10.1016/j.scitotenv.2008.09.037
31. Siedel H (2010) Historic building stones and flooding: changes of physical properties due to water saturation. J Perform Constructed Facil 24:452–461. https://doi.org/10.1061/(asce)cf.1943-5509.0000066
32. Török Á, Přikryl R (2010) Current methods and future trends in testing, durability analyses and provenance studies of natural stones used in historical monuments. Eng Geol 115:139–142. https://doi.org/10.1016/j.enggeo.2010.07.003
33. Ozga I, Bonazza A, Lyazidi SA et al (2013) Pollution impact on the ancient ramparts of the Moroccan city Salé. J Cult Heritage 14:S25–S33. https://doi.org/10.1016/j.culher.2012.10.018
34. Modestou S, Theodoridou M, Ioannou I (2015) Micro-destructive mapping of the salt crystallization front in limestone. Eng Geol 193:337–347. https://doi.org/10.1016/j.enggeo.2015.05.008
35. Forster AM, Carter K, Banfill PFG, Kayan B (2011) Green maintenance for historic masonry buildings: an emerging concept. Build Res Inf 39:654–664. https://doi.org/10.1080/09613218.2011.621345

36. Theodoridou M, Harbottle M (2020) Geological construction materials and structures—geoheal project—project ID 745891
37. Nardi C De, Theodoridou M, Sim P et al (2019) Self-healing lime-based mortars using biological mechanisms and microvascular networks. In: 5th Historic Mortars conference, Pamplona, Spain, 19–21 June 2019
38. Theodoridou M, Harbottle M (2021) Biological self-healing for natural stone in construction. In: Proceedings of the international symposium CONSOLIDATION 2021
39. Le Métayer-Levrel G, Castanier S, Orial G et al (1999) Applications of bacterial carbonatogenesis to the protection and regeneration of limestones in buildings and historic patrimony. Sediment Geol 126:25–34. https://doi.org/10.1016/S0037-0738(99)00029-9
40. Tiano P, Cantisani E, Sutherland I, Paget JM (2006) Biomediated reinforcement of weathered calcareous stones. J Cult Heritage 7:49–55. https://doi.org/10.1016/j.culher.2005.10.003
41. Harbottle MJ, Al-Tabbaa A (2008) Degradation of 2-chlorobenzoic acid in stabilised/solidified soil systems. Int Biodeterior Biodegradation 61:173–181. https://doi.org/10.1016/j.ibiod.2007.07.002
42. Jroundi F, Gómez-Suaga P, Jimenez-Lopez C et al (2012) Stone-isolated carbonatogenic bacteria as inoculants in bioconsolidation treatments for historical limestone. Sci Total Environ 425:89–98. https://doi.org/10.1016/j.scitotenv.2012.02.059
43. Dhami NK, Sudhakara Reddy M, Mukherjee A (2014) Application of calcifying bacteria for remediation of stones and cultural heritages. Front Microbiol 5:304
44. Perito B, Marvasi M, Barabesi C et al (2014) A Bacillus subtilis cell fraction (BCF) inducing calcium carbonate precipitation: biotechnological perspectives for monumental stone reinforcement. J Cult Heritage 15:345–351. https://doi.org/10.1016/j.culher.2013.10.001
45. De Muynck W, De Belie N, Verstraete W (2010) Microbial carbonate precipitation in construction materials: a review. Ecol Eng 36:118–136. https://doi.org/10.1016/j.ecoleng.2009.02.006
46. Botusharova S, Gardner D, Harbottle M (2020) Augmenting microbially induced carbonate precipitation of soil with the capability to self-heal. J Geotech Geoenvironmental Eng 146:04020010. https://doi.org/10.1061/(asce)gt.1943-5606.0002214
47. Jonkers HM, Thijssen A, Muyzer G et al (2010) Application of bacteria as self-healing agent for the development of sustainable concrete. Ecol Eng 36:230–235. https://doi.org/10.1016/j.ecoleng.2008.12.036
48. Joseph C, Lark R, Isaacs B et al (2010) Experimental investigation of adhesive-based self-healing of cementitious materials. Mag Concr Res 62:831–843. https://doi.org/10.1680/macr.2010.62.11.831
49. Tan L, Reeksting B, Ferrandiz-Mas V et al (2020) Effect of carbonation on bacteria-based self-healing of cementitious composites. Constr Build Mater 257:119501. https://doi.org/10.1016/j.conbuildmat.2020.119501
50. HBBE—biotechnology in the built environment. http://bbe.ac.uk/. Accessed 17 May 2021
51. Bridgens B (2018) How biotechnology can transform delivery and operation of the built environment. Proc Inst Civ Eng 173:13
52. Ng SHY, Lourenço SDN (2016) Conditions to induce water repellency in soils with dimethyldichlorosilane. Geotechnique 66:441–444. https://doi.org/10.1680/jgeot.15.T.025
53. Lourenço SDN, Saulick Y, Zheng S et al (2018) Soil wettability in ground engineering: fundamentals, methods, and applications. Acta Geotech 13:1–14
54. Lin H, Lourenço SDN (2020) Physical degradation of hydrophobized sands. Powder Technol 367:740–750. https://doi.org/10.1016/j.powtec.2020.04.015